HORMONE RESEARCH I

M. Norvell and T. Shellenberger, editors

CONTENTS

SPECIAL ISSUE ON HORMONAL RESEARCH

Copyright © by

HEMISPHERE PUBLISHING CORPORATION
1025 Vermont Avenue, N.W., Washington, D.C. 20005

INTRODUCTION

The first annual NCTR hormone research symposium was held at the National Center for Toxicological Research, Jefferson, Arkansas in November 1974.

The purposes of that symposium were to combine a nucleus of competent scientists engaged in various estrogen research activities, to provide a forum for the exchange of ideas between these scientists and Food and Drug Administration (FDA) personnel participating in an estrogen research program currently underway at NCTR, to assist in establishing realistic and meaningful research goals within the scope of this program, and to suggest specific courses of action to most efficiently utilize existing facilities and monetary and personnel resources in attaining those goals.

The symposium was highly successful in that it accomplished the stated objectives. Its success was due to the enthusiastic and conscientious participation of the invited speakers, other FDA guests, the NCTR staff, and FDA personnel from the Washington, D.C. area. The excellent quality of the submitted papers and the active participation of a number of the attendees during the discussion sessions resulted in a significant contribution to the field of estrogen research.

Michael J. Norvell
Chairman of the NCTR Hormone Research Symposia

Thomas E. Shellenberger
Director of the NCTR Hormone Research Program

GUEST EDITORS

DISCUSSION PARTICIPANTS

D. A. CASCIANO
Mutagenesis Division, National Center for
Toxicological Research, Jefferson, Arkansas

S. R. CERNOSEK
Biochemistry Department, University of Arkansas
for Medical Sciences, Little Rock, Arkansas

C. L. CHEN
Department of Physiological Sciences, College
of Veterinary Medicine, Kansas State University,
Manhattan, Kansas

J. H. CLARK
Department of Cell Biology, Baylor College
of Medicine, Houston, Texas

L. L. DOYLE
Department of Obstetrics and Gynecology,
University of Arkansas for Medical Sciences,
Little Rock, Arkansas

L. J. FISCHER
Department of Pharmacology, The Toxicology
Center, University of Iowa, Iowa City, Iowa

R. W. GILLETTE
Meloy Laboratories, Inc., Springfield, Virginia

D. L. GREENMAN
Chronic Studies Division, National Center for
Toxicological Research, Jefferson, Arkansas

H. D. HAFS
Department of Dairy Science, Michigan State
University, East Lansing, Michigan

D. M. HENRICKS
Department of Food Science, Clemson University,
Clemson, South Carolina

W. C. HOBSON
The International Center of Environmental
Safety, Albany Medical College, Holloman
Air Force Base, New Mexico

C. C. KALTENBACH
Division of Animal Science, University of
Wyoming, Laramie, Wyoming

D. MEDINA
Department of Cell Biology, Baylor College
of Medicine, Houston, Texas

J. J. OLESON
Office of the Director, National Center
for Toxicological Research, Jefferson,
Arkansas

W. D. PRICE
Bureau of Veterinary Medicine, Food and
Drug Administration, Washington, D. C.

T. E. SHELLENBERGER
Division of Molecular Biology, National
Center for Toxicological Research,
Jefferson, Arkansas

R. A. SHENEFELT
Division of Teratology, National Center
for Toxicological Research, Jefferson,
Arkansas

G. L. WOLFF
Mutagenesis Division, National Center
for Toxicological Research, Jefferson,
Arkansas

C. ZERVOS
Office of the Commissioner, Food and Drug
Administration, Washington, D. C.

MOUSE MAMMARY TUMOR VIRUS AS A MODEL FOR VIRAL CARCINOGENESIS

Ronald W. Gillette

Meloy Laboratories, Inc., Springfield, Virginia

Evidence is presented suggesting that genetic control may be the limiting factor in mammary tumor virus (MTV) positive mammary oncogenesis. Studies with hybrid mice involving high and low MTV-expressing murine strains suggest that expression of MTV in milk may not be as important in tumorigenesis as previously thought. The usefulness of the murine MTV model in the study of mammary malignancy is discussed.

INTRODUCTION

Introduction of mammary tumor virus-positive (MTV+) mammary adenocarcinomas requires the interaction of three distinct factors, presence of MTV, hormones, and genetic background. Absence of any one of these factors results in greatly decreased numbers of mammary tumors. The influences of each component on mammary tumorigenesis is discussed in some detail.

RESULTS AND DISCUSSION

Presence of MTV

MTV+ adenocarcinomas typically contain large numbers of mature type-B particles. Such particles consist of an excentric nucleocapsid surrounded by a glycoprotein-containing envelope. Type-B particles are thought to be generated from intracytoplasmic A particles. A particles appear in the cytoplasm and migrate to the cell surface where the immature B particle forms prior to being shed from the cell surface. Immature B particles typically have "spikes" protruding from the virus envelope.

Subgroups of MTV are thought to differ antigenically depending on the murine source of the virus. Table 1 lists the subtypes of MTV present in commonly used strains of mice. MTV-S of C3H represents the standard Bittner "milk factor." All vary to some degree in their oncogenicity. The percentage of tumors induced in a given strain of mice may also depend on the presence of MTV-N, the nodule-inducing virus.

Hormones

MTV+ tumors vary somewhat in their sensitivity to steroid hormone levels. The RIII mammary adenocarcinomas generally exhibit the greatest response to hormone levels. Such tumors are characterized by multiple

---

Requests for reprints should be sent to Ronald W. Gillette, Meloy Laboratories, Inc., 6715 Electronic Drive, Springfield, Virginia 22151.

Journal of Toxicology and Environmental Health, 1:545–550, 1976

TABLE 1. Types of Murine MTV

MTV	Mouse strain	Oncogenicity
S	C3H	high
P	RIII	high
O	BALB/c	weak
D	DD	weak
A	A	weak (??)
N	all (??)	very weak

episodes of remission that depend on the physiologic status of the host. C3H tumors are much less sensitive to hormones.

The relationship between parity and tumorigenesis is shown in Table 2. In our colony approximately 95% of all RIII females that reach third parity develop MTV+ adenocarcinomas. Virgin females held under similar conditions produce only about 20% tumors. Males exhibit occasional breast tumors. Similar results were obtained using C3H strain mice.

MTV+ mammary adenocarcinomas have also been studied *in vitro*. Although requirements for prolactin, cortisone, and insulin have been established, this approach has proved highly unsatisfactory with regard to virus production. Improved virus expression in established lines of MTV+ tumor cells has recently been achieved by treating the cells with dexamethasone. There also appears to be a role for cyclic AMP and/or GMP in the *in vitro* expression of MTV.

Genetic Background

Perhaps the limiting factor in MTV tumor expression is the genetic makeup of the host. Recent evidence stresses the great importance of inherited repressor genes in controlling the expression of cellular MTV information that leads to tumorigenesis.

Table 3 illustrates a number of MTV+ and MTV− mouse strains commonly in laboratory use. As can be seen the latency of mammary tumors in the negative strains is much greater than in positive strains.

TABLE 2. Relationship of MTV+ Tumorigenesis and Pregnancy

Strain	Status	Percent tumors
RIII, female	> 3rd parity	95
RIII, female	virgin	20
RIII, male		3
C3H, female	> 3rd parity	90
C3H, female	virgin	10
C3H, male		< 1

TABLE 3. MTV Tumor Induction and Latency in Various Inbred Strains of Mice

	Percent tumors	Latency (months)
MTV+		
RIII	95	7–9
C3H/He	90	8–12
DD	85	12–14
GR	98	5–7
C3HA^{vy}	98	4–6
MTV−		
NIH Swiss	2	14–18
C57B1	<1	>24
C3H/Wi	5	15–18

An approach to studying the importance of repressor genes has been the study of expression of MTV antigens in milk samples from hybrid mice produced by reciprocal mating of individuals from high or low MTV expressor mouse strains. It was found that high expressor strains produced significant amounts of virus at the first parity; while negative strains eventually produce MTV milk antigen, they do so only after several pregnancies. Surprisingly, material influence in F_1 crosses of high and low expressor strains was minimal.

Tumor incidence and latency in hybrid females are compared in Table 4. As can be seen crosses involving high and low expressors produced fewer tumors than either crosses involving two high expressor strains or high expressor parental strain mice. Factors such as litter size, average numbers of litters produced, or daily variation in MTV antigen expressed in milk did not influence the results. In addition, forced breeding was found to be without effect on the expression of viral antigen in milk in that donors that were force bred to third parity produced viral antigen at levels comparable to first

TABLE 4. Tumor Incidence and Latency in F_1 Females

Mating	No. tumors/ total	Percent	Mean latency (months)
C3H X Swiss	2/14	14	8
Swiss X C3H	2/16	112	7
C3H X BALB/c	4/11	36	9
BALB/c X C3H	10/16	62	7
C3H X DD	2/14	14	7
DD X C3H	7/15	47	7
C3H X RIII	13/15	87	9
RIII X C3H	15/15	100	8
C3H X C3H	10/15	67	7
RIII X RIII	13/15	87	7

parity, normally bred females. It was concluded that expression of MTV in milk may not be as important in the transmission of MTV as previously thought. On the other hand, genetic factors may be the limiting influence in control of mammary tumor induction.

Support for the concept of genetic control of MTV tumorigenesis derives from two additional sources. Varmus and co-workers (1972) and Scolnick and his co-workers (1974) have demonstrated that the amount of MTV information in GR mice (high expressor) is about equal to that present in C57B1 (very low expressor) mice. In addition, E. M. Scolnick and W. Parks (personal communication) have found equivalent amounts of MTV information in cells from fetal and adult RIII donors. Since fetal information was present before suckling, the conclusion may be drawn that transmission of the virus occurred with the germinal cells.

CONCLUSION

MTV+ mammary tumor induction represents an excellent model for the study of viral oncogenesis. MTV infection is life-long and with the appropriate genetic background results in virtually 100% mammary tumors if adequate hormonal components are provided. In addition, the murine MTV model possesses many characteristics in common with the human disease and has proved to be a valuable tool in the study of human mammary malignancy.

QUESTIONS AND ANSWERS

T. E. Shellenberger: Would you comment on the possible influence of B-type particles in human mammary tissues and would you also discuss horizontal (not reproducing or nursing) transmission of MTV virus in C3H mice as well as other species?

R. W. Gillette: Moore et al. (1971) have published some recent EM work that suggests that MTV type-B particlelike structures do exist in human milk samples. Shlom et al. (1971), using DNA hybridization studies, have claimed they have found MTV-like information in the DNA from human material. I do not imply that horizontal (cage-to-cage) MTV transmission does not occur. It may occur in a manner similar to a "super" infection, although it is a rare event. I feel that tolerance in the animal may have a lot to do with its susceptibility to MTV material when horizontally transmitted. However, the virus is normally transmitted only during mating or nursing.

D. A. Casciano: Are the data good enough to rule out incomplete virus vs. protecting repressor genes?

R. W. Gillette: No, we do not have infectivity information on type-B virus due to the lack of good assay procedures in this area. This lack of information is a liability in MTV work.

D. Medina: You stated that horizontal transmission may occur. If horizontal transmission (cage-to-cage) does occur (and there is very little evidence for it), what then is the significance of the findings of Varmus et al. (1972) and Scolnick et al. (1974) on the low levels of MTV-like DNA in BALB/c and C57 mice?

R. W. Gillette: As far as I know, MTV is only transmitted by breeding or nursing.

D. Medina: Has this been tested under rigorous conditions, such as by taking C57 or BALB/c mice, isolating them in a barrier, and measuring the titer to MTV antigens?

R. W. Gillette: Not to my knowledge.

J. J. Oleson: John Bittner indicated to me many years ago that MTV virus is found in the saliva of MTV mice.

D. Medina: Some published data suggest that cage-to-cage horizontal transmission may occur. Blair (1974), at Berkeley, caged virus-positive and virus-negative mice together and then separated them. When bloodborne MTV antigens were tested in the titer of the "mixed" animals in adjacent cages it rose; when she separated the cages of mice, the titer eventually fell. While these observations only suggest that MTV virus may be transmitted across cages (airborne), it does indicate the need to conduct a rigorous experiment in this area.

R. W. Gillette: Mammary tumor virus in C3H mice seems to be similar to type-C virus in other strains of animals; that is, with rigorous testing one can find it in all strains. It is much easier to find MTV+ than MTV− mice. It is really a matter of degree of infectivity and not so much an all-or-none effect. By exhaustively using RIA, immunofluorescence of B lymphocytes, and DNA hybridization techniques, MTV virus can be found in all strains.

R. A. Shenefelt: You are saying that all mice seem to have MTV virus. Is there no absolute test for negative mice?

R. W. Gillette: I think one must look for repressor genes. Eventually, all animals will express the presence of the virus, but it will take longer for the more resistant strains to do so than the more susceptible strains. Again, stress is important, given the effects of breeding on the incidence of mammary tumors in mice.

Unknown: Is there any evidence that separate genes exist in the mouse?

R. W. Gillette: To my knowledge, the appropriate genetic studies have not been done in the mouse.

J. H. Clark: Have enough studies been conducted to determine the effects of one or more hormones on the susceptibility of mice to MTV virus?

R. W. Gillette: Not many studies have been done in this area. Some workers feel that prolactin is required for the expression of MTV virus. If animals are given multiple pituitary grafts, the incidence of tumors is rapidly increased to a very high incidence.

D. Medina: Have you ever looked for MTV antigen in rat milk?

R. W. Gillette: No.

G. L. Wolff: Has the hormone balance in the C3H A^{vy} mice been investigated?

R. W. Gillette: Not to my knowledge. It is interesting to note that the C3H^{vy}fB mice express tumors but do not express the virus in the milk.

REFERENCES

Blair, P. 1974. Immunologic evidence for horizontal transmission of MTV. *J. Immunol.* 113:1446–1449.

Moore, D. H., Charney, J., Kramarsky, B., Lasfarguos, E. Y., Sarkar, N. H., Brennan, M. J., Burrows, J. H., Sirsat, S. M., Paymaster, J. C. and Vaida, A. B. 1971. Search for a human breast cancer virus. *Nature (London)* 229:611–614.

Schlom, J., Spiegelman, S. and Moore, D. 1971. RNA-dependent DNA polymerase activity in virus-like particles isolated from human milk. *Nature (London)* 231:97–100.

Scolnick, E. M., Parks, W., Kawakomi, T., Kohne, D., Okahe, H., Gilden, R. and Hotonaka, M. 1974. Primate and murine type-C viral nucleic acid association kinetics: Analysis of model systems and natural tissues. *J. Virol.* 13:363–369.

Varmus, H. E., Bishop, J. M., Nowinski, R. C. and Sarkar, N. H. 1972. Mammary tumor virus specific nucleotide sequences in mouse DNA. *Nature New Biol.* 238:189–191.

Received June 4, 1975
Accepted June 27, 1975

MAMMARY GLAND AS A MORPHOLOGICAL END POINT IN CARCINOGENESIS STUDIES

Daniel Medina

Department of Cell Biology, Baylor College of Medicine,
Houston, Texas

The murine mammary tumor system is characterized by the presence of intermediate preneoplastic mammary cell populations. The intermediate mammary cell populations are characterized as either alveolar hyperplasias or ductal hyperplasias. The most frequently encountered preneoplastic cell population is the hyperplastic alveolar nodule, which has been extensively characterized with respect to its biological and hormonal properties. While the alveolar hyperplasias are seen predominantly in mice infected with the mammary tumor virus, ductal hyperplasias are seen primarily in mice treated with chemical carcinogens. Both types of lesions can serve as morphological end points for carcinogenesis studies in the mammary gland.

INTRODUCTION

The established dogma in cancer biology developed over the past 50 yr has considered the appearance of a palpable tumor as the desirable, if not only, end point in carcinogenesis studies. However, studies over the past 15 yr, particularly by DeOme and co-workers (DeOme and Medina, 1969; DeOme and Nandi, 1966; Medina, 1974; Nandi, 1966), have led to a model of mammary tumorigenesis in rodents that acquires special significance in any discussion of studies on end points in carcinogenesis. The scheme depicted below represents the present state of our knowledge regarding the sequence of events in mouse mammary tumorigenesis.

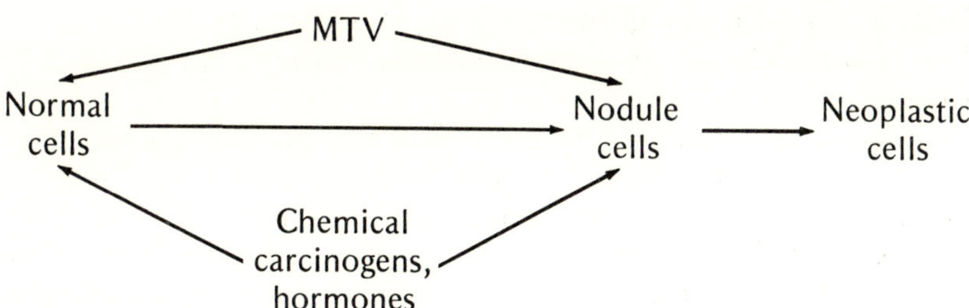

The model suggests several concepts that are important in any discussion

This work was supported by USPHS research grants CA-11944 and CA-10893.
We gratefully acknowledge the technical assistance of Betsey Perry and Joy Ewing.
Requests for reprints should be sent to Daniel Medina, Department of Cell Biology, Baylor College of Medicine, Houston, Texas 77025.

Journal of Toxicology and Environmental Health, 1:551–560, 1976

of mammary tumorigenesis. At least two sequential transformations occur in the process of mammary tumorigenesis. First, normal cells change to nodule cells (nodule transformation). This event leads to the appearance of subpopulations of mammary cells (nodules) that appear as either alveolar or ductal hyperplasias. They can be identified by their occurrence in nonpregnant, nonlactating female mice. These nodules are considered pre-neoplastic lesions since they give rise to neoplasms more frequently and with shorter latent periods than normal mammary tissues (DeOme et al., 1959; Medina, 1973).

The second transformation is the neoplastic transformation (i.e., the change from nodule to neoplastic cells). Although the model does not rule out a one-step transformation from normal to neoplastic, the usual sequence of progression includes at least both transformations.

In the course of this discussion, the following questions will be considered: What are the types of nodules seen in the mammary glands of mice and what are their properties? What oncogenic agents are capable of inducing the appearance of nodules? Can the incidence or presence of mammary nodules form the basis for an assay for carcinogenic agents?

RESULTS AND DISCUSSION

The most frequently encountered nodule is termed the hyperplastic alveolar nodule (HAN). These HAN, which resemble the normal mammary lobules of pregnant mice, can be identified as discrete foci of alveolar hyperplasias in the mammary glands of nonpregnant, nonlactating female mice. Some characteristic properties of HAN are the following:

Upon transplantation into the cleared mammary fat pads of 3-wk-old syngeneic young female mice, they grow and fill the fat pad with alveolar hyperplasia (DeOme et al., 1959). Once the entire fat pad is filled (at 10 wk), net growth ceases. At some time thereafter, months after transplantation, mammary tumors evolve from the nodule outgrowths.

The frequency of tumor formation depends on the nodule line used and the strain in which the nodules arise (Table 1). Thus, nodules occurring in the mammary tumor virus-positive (MTV+) C3H mice produce a high incidence of tumors on transplantation (Blair and DeOme, 1961; DeOme et al., 1959), whereas nodules arising in MTV− C3Hf mice produce a low incidence of tumors (Blair and DeOme, 1961; Medina et al., 1970). HAN, which occur in 7, 12-dimethylbenz-anthracene (DMBA)-treated Lewis rats, are also preneoplastic (Beuving, 1968). Additionally, the tumor potential varies with the individual nodule line. A nodule outgrowth is established by serially transplanting nodule outgrowths at periodic intervals in the fat pads of young syngeneic mice. These lines are *in vivo* analogues to *in vitro* cell lines.

TABLE 1. Tumor-Producing Capabilities of Hyperplastic Alveolar Nodules

Strain	Transplant[a]	No. tumors/ No. transplants	Percent	Mean time of tumor appearance (wk)
C3H	HAN	11/15	73	20.9
C3H	NMG	2/15	13	30.5
C3H	HAN	20/29	69	20.4
C3H	NMG	3/32	9	26.0
C3Hf	HAN	2/15	13	N.D.[b]
C3Hf	HAN	1/14	7	45.0
C3H/2	HAN	1/17	2	26.0
BALB/c	HAN	5/5	100	N.D.
BALB/cfC3H	HAN	12/15	80	30.3
Lewis rat	HAN	8/37	22	N.D.
Lewis rat	NMG	0/37	0	N.D.

[a]HAN, hyperplastic alveolar nodule; NMG, normal mammary gland.
[b]N.D., no data.

Thus, in two MTV− BALB/c nodule lines (Table 2), one line (D1) produces only 8% mammary tumors by 16 months, whereas another line (D2) produces 50% mammary tumors by 12 months.

Most HAN arising in either MTV+ C3H mice or MTV− BALB/c mice are independent of ovarian and adrenal hormones, but not anterior pituitary hormones, for growth and tumorigenesis (Medina, 1973). Conversely, exogenously administered prolactin enhances the tumorigenic transformation in ovariectomized–adrenalectomized mice (Yanai and Nagasawa, 1971).

Most HAN have very low levels of either nuclear localization of estradiol receptor or cytoplasmic/nuclear localization of estradiol receptor (D. Medina, unpublished). Thus, MTV-induced and spontaneously arising nodules and tumors are ovarian and adrenal independent by several criteria and contrast with chemical carcinogen-induced mammary tumors in the rat.

HAN are fat-pad dependent and cannot grow in subcutaneous tissues or other ectopic sites.

HAN are capable of infinite propagation (Daniel et al., 1968). HAN

TABLE 2. Tumor-Producing Capabilities of BALB/c Nodule Outgrowth Lines D1 and D2

Nodule line	No. tumors/ No. transplants	Percent	Mean latent period of tumors (days)	Mean age of mice without tumors (days)
D1	5/62	8.1	410	495
D2	45/89	50.6	240	304

can be serially transplanted in the fat pads of young syngeneic mice indefinitely, whereas normal tissues die out after five transplant generations (60 wk).

The above criteria characterize the HAN as an intermediate stage in the progression from normal to neoplasia and provide a possible end point for carcinogenesis studies.

The second most frequent lesions encountered are ductal hyperplasias. These lesions appear as atypical proliferations of mammary ducts with abundant intraluminal epithelial proliferations in the nonpregnant, non-lactating animal. The hyperplastic mammary ducts are often capped by end buds. Histologically, they appear as atypical lesions that exhibit varying degrees of dysplastic growth. Questions related to their preneoplastic nature and hormonal dependencies are now under investigation.

Another type of ductal lesion, the "plague," has been described by Foulds (1949, 1956) in certain strains of mice of European origin. Plagues are hormone-dependent neoplasms that occur in MTV+ GR or RIII mice. Their progression into hormone-independent neoplasms has also been extensively described by Foulds. These lesions are not found in common American-derived strains of mice, such as C3H, DBA, and BALB/c. Their relationship to the previously described ductal hyperplasias, which occur in C57BL, DBA, and BALB/c mice, is under investigation.

Finally, lesions described as either inflammatory, inflammatory reactive, or keratinizing nodules occur in a variety of strains (Huseby and Bittner, 1946; Medina, 1974). These have very low or no tumor potential because, on transplantation, they give rise to stunted growth or ductal growth. These lesions are incidental responses to carcinogens and are considered dead-end lesions.

The second question of importance pertains to the nature of the oncogenic agents capable of inducing the various alveolar and ductal lesions. In Table 3, note that in MTV+ BALB/c and C57BL mice the primary lesion induced is the HAN in large quantities (Medina et al., 1970; Nandi et al., 1972). In MTV+ BALB/c, approximately 25 HAN/mouse were detected in 100% of the mice. In contrast, MTV− BALB/c mice treated with 6.0 mg DMBA and pituitary isografts contained very few HAN and a few ductal hyperplasias, but a relatively large percentage of mammary tumors (68%). Ductal hyperplasias (DH) are frequent in urethan-treated C57BL and (C57BL X DBAf)F_1 mice, but mammary tumors are infrequent. However, DMBA-treated virgin BALB/c mice contain a relatively high percentage of ductal hyperplasias and 33% mammary tumors. These results suggest that intermediate nodule lesions (alveolar and ductal) are present in carcinogen-treated mice but not in large quantities as they are in MTV mice. Also the majority of lesions are ductal, rather than alveolar hyperplasias. The preneoplastic potential and hormonal properties of ductal hyperplasias are unclear; however, recent studies have shown that

TABLE 3. Noduligenesis in BALB/c and C57BL Mice

Group	Strain	Carcinogen	No. of mice with HAN/ Total no. of mice	Percent	Mean no. of HAN/ HAN-bearing mice	No. of mice with DH/ Total no. of mice	Percent	Mean no. of DH/ DH-bearing mice	Percent tumors
1	BALB/c	MTV[a]	13/15	87	25.0	0	0	0	0
2	BALB/c	DMBA (6.0 mg)[a]	6/20	30	3.3	4/20	20	1.0	68
3	BALB/c	DMBA (4.0 mg)[b,c]	0/19	0	0	10/19	53	4.9	33
4	C57BL	MTV	12/15	80	4.2[d]	N.D.[e]	—	—	N.D.
5	C57BL	DMBA (6.0 mg)[a]	4/23	17	1.8	1/23	4	1.0	33
6	C57BL	Urethan (in water)[a]	5/10	50	5.2	6/10	60	1.8	0
7	C57BL	Urethan (0.01%)[a]	1/9	11	1.0	6/9	67	1.7	0
8	(C57 × DBA/f)F$_1$	Urethan (200 mg)[a]	0/13	0	0	10/13	77	4.3	15

[a]Mice received a single pituitary isograft under the kidney capsule for 3 months.
[b]Virgin mice.
[c]All groups were terminated at 7–8 months except groups 3 and 7, which were terminated at 10 and 12 months, respectively.
[d]Calculated as mean number of HAN/HAN-bearing mammary gland.
[e]N.D., no data.

the few HAN occurring in DMBA-treated, pituitary isograft-bearing mice have very high tumor potentials. Thus, the presence of one HAN or DH per mouse is enough to ensure a relatively high tumor incidence (D. Medina, unpublished).

Mammary tumor viruses in the mouse and chemical carcinogens, such as dimethylbenzanthracene, methylcholanthrene, benzpyrene, urethane, and acetylaminofluorene, have been established for a long time as mammary carcinogens in the mouse and the rat. The question of the carcinogenicity of estrogens for mammary tissue free of MTV has not been clearly answered. In a variety of rat strains, it is well established that prolonged administration of estrogens (estrone pellets) will produce a high incidence of mammary adenocarcinomas (Cutts and Noble, 1964a, b). In MTV+ C3H mice, diethylstilbestrol given at low doses for prolonged times in the food will enhance mammary tumorigenesis (Gass et al., 1964). Whether this effect is on the noduligenic transformation or the tumorigenic transformation has not been elucidated, but the effect is not surprising since many hormonal regimes enhance the tumor potential of established HAN (Medina, 1973). The critical question is the effect of estrogens in MTV— mice.

One of the few studies of interest is the effect of perinatal administration of estradiol-17β on subsequent noduligenesis in MTV— BALB/c mice (Warner and Warner, 1975). Mice given 50 μg estradiol-17β for the first 5 days of life and left untreated thereafter did not have a significantly enhanced mammary lesion (dysplasias and neoplasms) incidence over untreated controls. However, such neonatally treated mice developed more mammary dysplasias than normal mice when both groups were exposed to DMBA at 4 wk, 8 wk, or 6 months of age. This type of experiment stresses the importance of the interplay of various carcinogenic and noncarcinogenic agents in the induction of preneoplastic and neoplastic lesions.

The presence of hyperplastic mammary lesions in the sequence of mammary tumor progression serves, under special conditions, as a useful basis for a more rapid *in vivo* assay than the appearance of tumors. The presence of a biologically active mammary tumor virus can be detected by the induction of HAN nodules in hormone-primed mice within 11–14 wk. Tumor development takes 6–8 months. The assay is reliable and can be semiquantitated (Nandi et al., 1971). However, this assay is not completely reliable if one wishes to demonstrate the carcinogenicity of a chemical since mammary tumors may arise in the absence of detectable ductal or alveolar hyperplasias (Medina, 1974). A similar phenomena occurs in the chemical-carcinogen-treated rat under special conditions (Sinha and Dao, 1974). Thus, tumor appearance may be more reliable than a nodule assay for chemical carcinogens. The tumors that arise in chemical-carcinogen-treated mice are both adenocarcinomas and adenoacanthomas, compared with only adenocarcinomas in MTV-infected mice.

Since the sequence in murine mammary tumorigenesis involves a pre-neoplastic stage, one might ask whether nodules that arise in the absence of overt oncogenic stimuli are susceptible to further carcinogenic agents. Table 4 documents that nodule outgrowth lines derived from HAN in MTV-free mice are responsive to a variety of viral, chemical, and physical oncogenic agents. In a virus-free system, it is likely that the response of nodule tissue to suspected oncogens may provide additional and valuable information.

Any approach that attempts to evaluate the significance of suspected mammary carcinogens must consider the following points: (1) the ability of the agents to induce various types of preneoplastic nodules, as well as neoplasms; (2) the effects of these agents on the nodule to tumor transformation; and (3) the effect of these agents on enhancing the susceptibility of the target organ to subsequent exposure of known and unknown carcinogens. It is possible that estrogens, particularly when given perinatally, fall in the third category. A corollary of this third alternative is the possibility that known carcinogens may modify the response of target organs to the hormonal milieu of the host. Table 5 shows several examples of this phenomenon where exposure to methylcholanthrene and dimethylbenzanthrene increased the susceptibility of the mammary tissues for tumor induction in an otherwise innocent hormonal milieu.

CONCLUSION

If the rodent model is applicable to man, it is incumbent on those interested in detecting potential human mammary carcinogens to evaluate such agents for both their noduligenic and tumorigenic potentials under

TABLE 4. Effect of Viral, Chemical, and Physical Carcinogens on Tumor Incidence in D Series Nodule Outgrowths

Outgrowth line	Carcinogen	Dose	No. tumors/ No. transplants	Percent tumors	Mean latent period (days)
D1	—	—	10/250	4	404
D1	MTV	—	91/122	75	266
D1	MCA	1.5 mg	30/52	58	230
D1	DMBA	1.5 mg	80/119	67	207
D1	Urethan	200 mg	73/93	78	209
D1	Irradiation	450 R	16/69	23	254
D1	Hormone stimulation[a]	—	4/58	7	371
D2	—	—	41/92	45	280
D2	MTV	—	43/53	81	182
D2	MCA	1.5 mg	18/20	90	230
D2	Irradiation	450 R	25/41	61	245
D2	Hormone stimulation	—	50/65	77	227

[a]Pituitary isografts.

TABLE 5. Effect of Hormonal Milieu of Host on the Tumor-Producing Capabilities of Chemical Carcinogen-Treated Nodule Outgrowth Line D1

Donor tissue	Hormonal milieu of recipient	No. tumors/ No. transplants	Percent	TE_{50} (days)[a]
—	Pituitary isograft	2/20	10	—
MCA	Virgin	14/148	10	—
MCA	Pituitary isograft	49/62	79	260
MCA	Neonatal E_2	10/26	38	—
MCA-1-one	Pituitary isograft	12/19	63	329
DMBA	Virgin	14/23	61	233
DMBA	Pituitary isograft	25/30	83	210
DMBA	Breeder	14/18	78	171
—	Breeder	1/24	4	—[b]

[a]Time for 50% of the transplants to produce tumors.
[b]Experiment terminated at 330 days.

the variety of conditions listed above to gain an accurate picture of their true potentials.

QUESTIONS AND ANSWERS

J. J. Oleson: To what extent would previous exposure to carcinogens affect test animals being exposed to other test substances?

D. Medina: At this time, we do not have sufficient data to determine the extent of pretest exposure interference. About all we can say now is that these things should be accounted for as much as possible.

R. A. Shenefelt: Could conditions leading to ductal hyperplasias in neonatal mice be controlled so as to mimic mammary carcinoma in man?

D. Medina: Information concerning ductal hyperplasias has only been accumulated during the past year. We are currently investigating this possibility. Only recently have we been able to obtain the required conditions for producing large amounts of ductal hyperplasias.

R. W. Gillette: Could the growth of the transplanted tumor be due to some kind of immunological phenomenon since the transplanted tumors grow only in the fat pad?

D. Medina: That is a consideration; however, in all the tests conducted these tissues have been shown to have a very weak immunogenic potential, either by *in vivo* assays where it cannot be detected at all or by *in vitro* assays where it is detected only to a limited extent. Second, these are all done in syngeneic animals in which immune reactions are not histologically detectable.

W. D. Price: As far as you can tell, are all mammary tumors observed in mice in all cages the same?

D. Medina: In the virus-positive animal they tend to be the same; that is, adenocarcinomas are not well differentiated. In a carcinogen-treated animal, one sees a high incidence of adenoacanthomas with squamous metaplasia.

W. D. Price: Are these mouse tumors useful as models for studying mammary carcinomas in man?

D. Medina: In MTV+ mice, most carcinomas result from alveolar hyperplasias, whereas in man most carcinomas arise from ductal tissues. However, chemical-carcinogen-induced mammary tumors in mice arise from ductal tissues and resemble human carcinomas in many respects.

J. H. Clark: Neonatal rats subjected to estrogens become acyclic and the blood estrogen levels are known to be increased under these conditions. In your experiments, have you measured the blood levels of estrogens in the mice receiving the carcinogen to check out the possibility of a carcinogen-endogenous estrogen synergism?

D. Medina: Blood estrogen levels have not been checked in these mice yet.

T. E. Shellenberger: What is the dose level of estrogen used?

D. Medina: Fifty micrograms per day for the first 5 days.

H. D. Hafs: Could you do the same thing by delaying the estrogen treatment for 2 or 3 wk?

D. Medina: I do not know. In the early work by Bern and his group in reporting changes in the vagina, they have narrowed the changes resulting from estrogen neonatal treatment to the first 5 days.

J. H. Clark: Could it result from continuous exposure to estrogen and associated acyclic changes?

D. Medina: In these studies, estradiol does not seem to be a primary carcinogenic agent; rather, it is a permissive agent. We have not investigated DES under these conditions.

H. D. Hafs: Has anyone looked at estrogenized males?

D. Medina: Not in these types of experiments. Some work was done in rats in which the administration of estrogen was changed to low doses over a long period of time. Under these conditions, one sees a high level of estrogen and a high level of prolactin with a decrease in the incidence of mammary tumors.

REFERENCES

Beuving, L. 1968. Mammary tumor formation within outgrowths of transplanted hyperplastic nodules from carcinogen-treated rats. *J. Natl. Cancer Inst.* 40:1287–1291.

Blair, P. B. and DeOme, K. B. 1961. Mammary tumor development in transplanted hyperplastic alveolar nodules of the mouse. *Proc. Soc. Exp. Biol. Med.* 108:289–291.

Cutts, J. H. and Noble, R. L. 1964a. Estrone-induced mammary tumors in the rat. I. Induction and behavior of tumors. *Cancer Res.* 24:1116–1123.

Cutts, J. H. and Noble, R. L. 1964b. Estrone-induced mammary tumors in the rat. II. Effect of alterations in the hormonal environment on tumor induction, behavior, and growth. *Cancer Res.* 24:1124–1130.

Daniel, C. W., DeOme, K. B., Young, L. J. T., Blair, P. B. and Faulkin, L. J., Jr. 1968. The *in vivo* lifespan of normal and preneoplastic mouse mammary glands. A serial transplantation study. *Proc. Natl. Acad. Sci. U.S.A.* 61:52–60.

DeOme, K. B. and Medina, D. 1969. A new approach to mammary tumorigenesis in rodents. *Cancer* 24:1255-1258.

DeOme, K. B. and Nandi, S. 1966. The mammary-tumor system in mice: A brief review. In *Viruses-inducing cancer*, ed. W. J. Burdette, pp. 127-137. Salt Lake City: Univ. Utah Press.

DeOme, K. B., Faulkin, L. J., Jr., Bern, H. A. and Blair, P. B. 1959. Development of mammary tumors from hyperplastic alveolar nodules transplanted into gland-free mammary fat-pads of female C3H mice. *Cancer Res.* 19:515-520.

Foulds, L. 1949. Mammary tumors in hybrid mice: Growth and progression of spontaneous tumors. *Br. J. Cancer* 3:345-375.

Foulds, L. 1956. The histologic analysis of mammary tumors in mice. *J. Natl. Cancer Inst.* 17:701-801.

Gass, G. H., Coats, D. and Graham, N. 1964. Carcinogenic dose–response curve to oral diethyl-stilbestrol. *J. Natl. Cancer Inst.* 33:971-977.

Huseby, R. A. and Bittner, J. J. 1946. A comparative morphological study of the mammary glands with reference to the known factors influencing the development of mammary carcinoma in mice. *Cancer Res.* 6:240-255.

Medina, D. 1973. Preneoplastic lesions in mouse mammary tumorigenesis. *Methods Cancer Res.* 7:3-53.

Medina, D. 1974. Mammary tumorigenesis in chemical carcinogen-treated mice. I. Incidence in BALB/c and C57BL mice. *J. Natl. Cancer Inst.* 53:213-221.

Medina, D., DeOme, K. B. and Young, L. 1970. Tumor-producing capabilities of hyperplastic alveolar nodules in virgin and hormone-stimulated BALB/cfC3H and C3HF mice. *J. Natl. Cancer Inst.* 44:167-174.

Nandi, S. 1966. Interactions among hormonal, viral and genetic factors in mouse mammary tumorigenesis. *Can. Cancer Conf.* 6:69-81.

Nandi, S., Haslam, S., Helmich, C. and Ritter, R. I. 1971. Nodule bioassay of mouse MTV: Modifications and quantitative considerations. *J. Natl. Cancer Inst.* 46:1309-1315.

Nandi, S., Haslam, S. and Helmich, C. 1972. Mechanisms of resistance to mammary tumor development in C57BL and I strains of mice. I. Noduligenesis, tumorigenesis and characteristics of nodules and tumors. *J. Natl. Cancer Inst.* 48:1005-1012.

Sinha, D. and Dao, D. L. 1974. A direct mechanism of mammary carcinogenesis induced by 7,12-dimethylbenzanthracene. *J. Natl. Cancer Inst.* 53:841-846.

Warner, M. R. and Warner, R. L. 1975. Effects of exposure of neonate mice to 17β-estradiol on subsequent age-incidence and morphology of carcinogen-induced mammary dysplasias. *J. Natl. Cancer Inst.* 55:289-298.

Yanai, R. and Nagasawa, H. 1971. Enhancement of pituitary isografts of mammary hyperplastic nodules in adreno-ovariectomized mice. *J. Natl. Cancer Inst.* 46:1251-1255.

Received June 4, 1975
Accepted June 27, 1975

ESTROGEN–RECEPTOR BINDING: RELATIONSHIP TO ESTROGEN–INDUCED RESPONSES

J. H. Clark, E. J. Peck, Jr.

Department of Cell Biology, Baylor College of Medicine,
Houston, Texas

J. N. Anderson

Department of Pharmacology, Stanford University
Medical Center, Stanford, California

The relationship between nuclear binding of the receptor–estrogen complex $(R \cdot E)$ and estrogenic responses was examined. Nuclear-bound $R \cdot E$ was measured by the $[^3H]$estradiol exchange assay, which permits the quantification of $R \cdot E$ as a function of either endogenous or nonlabeled estrogen. The quantity of nuclear $R \cdot E$ fluctuated as a function of blood levels of estrogen during the estrous cycle in the rat. This indicates that the accumulation of the $R \cdot E$ complex by the nucleus of uterine cells may be of physiologic significance and under the control of endogenous estrogen.

Immature rats received a single injection of estradiol (E_2) or estriol (E_3) and the following parameters were measured: accumulation and retention of the estrogen receptor by the nucleus of uterine cells; incorporation of $[^{14}C]$glucose into CO_2, lipid, protein, and RNA; RNA polymerase activity; water imbibition; and increased dry weight. All early uterotropic responses (0–3 hr) that were measured were equally stimulated by E_2 and E_3. However, E_3 failed to stimulate true uterine growth (increased dry weight 24 hr after injection), whereas E_2 produced a significant stimulation of true uterine growth. These data suggest that the $R \cdot E$ complex is capable of stimulating early uterotropic events, regardless of which estrogen is present. To produce true uterine growth, however, the $R \cdot E$ complex must be retained in the nucleus for long periods of time.

The mechanism of action of nonsteroidal estrogen antagonists was also investigated. Nafoxidine, CI-628, or clomiphene was injected either alone or in combination with estradiol, and the uterine weight and cytoplasmic estrogen–receptor content were determined. All antagonists were clearly uterotropic and not antagonistic at 24 hr after injection. These data demonstrate that nonsteroidal estrogen antagonists are antagonists not because they compete for cytoplasmic estrogen receptors to produce a complex that is a poor stimulator of growth, but because they fail to stimulate the replenishment of the receptor.

This work was supported by USPHS grant HD-04985 and by American Cancer Society grant BC-92.

Requests for reprints should be sent to J. H. Clark, Department of Cell Biology, Baylor College of Medicine, Houston, Texas 77025.

Journal of Toxicology and Environmental Health, 1:561–586, 1976

INTRODUCTION

The interactions of the steroid hormones estrogen and progesterone with their respective target tissues appear to depend on their binding to specific macromolecules called receptors. The factors involved in steroid–receptor binding and a generalized scheme of the events that occur in a typical target cell are shown in Fig. 1. Steroid hormones (S) appear to enter all cells by passive diffusion but are retained only in their respective target organs (Gurpide and Welch, 1969; Jensen and DeSombre, 1972; Peck et al., 1973). This retention is due to the presence of macromolecules in the cytoplasm, receptors (R_c), that bind the steroid to form a receptor–steroid complex ($R_c \cdot S$), which moves to the nucleus (Jensen et al., 1968; Shyamala and Gorski, 1969). The appearance of the receptor–steroid complex in the nucleus ($R_n \cdot S$) is a specific process initiated by the respective sex steroid and not by other steroid hormones (Anderson et al., 1972a); it occurs under physiologic circumstances (Clark et al., 1973). The $R_n \cdot S$ complex appears to bind to specific sites on chromatin called acceptor sites (A). This binding interaction leads to synthetic events that result in the stimulation of new mRNA synthesis. The mRNA is translated into proteins as shown in Fig. 1; the resultant protein products are involved in regulation of cell growth and metabolism.

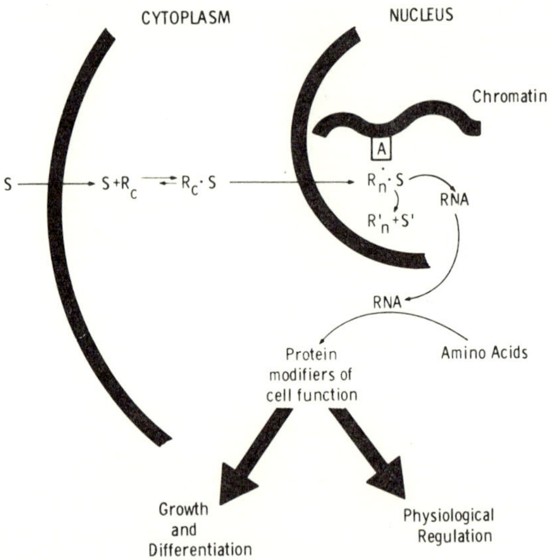

FIGURE 1. Current concept of the mechanism of steroid hormone action. See text for details and explanation. S, steroid; R_c, cytoplasmic receptor; $R_c \cdot S$, cytoplasmic receptor–steroid complex; $R_n \cdot S$, receptor–steroid complex in the nucleus; R'_n, modified (inactivated or degraded); S', metabolized steroid; RNA, ribonucleic acid.

Subsequent to the binding of $R_n \cdot S$ to A sites in the nucleus, the receptor and the steroid probably undergo metabolic changes to nonfunctional forms R_n' and S'.

We are interested specifically in the relationships that exist between receptor–estrogen binding and estrogen-induced uterine growth. The important uterotropic events that are involved in the stimulation of uterine growth appear to originate in the nuclei of uterine cells and probably arise for receptor–estrogen $(R \cdot E)$ interactions at the nuclear level.

RESULTS AND DISCUSSION

Determining the Concentration of Nuclear $R \cdot E$ Complex

The major impediment to the study of the relationship between the quantity of nuclear $R \cdot E$ and estrogen-induced responses under *in vivo* conditions has been the lack of an adequate assay for the measurement of $R \cdot E$. Previous methods for evaluating the quantity of nuclear-binding sites have utilized gel filtration or density-gradient ultracentrifugation of nuclear extracts solubilized from the uterine nuclear fraction with buffered 0.3–0.4 M KCl solutions. These procedures are laborious, difficult to quantitate, or require the injection of [^{3}H]estradiol for *in vivo* studies. We have developed a method for the determination of the concentration of $R \cdot E$ in the nuclear fraction of estrogen target tissues that circumvents these difficulties, thus permitting an evaluation of the concentration of specific estradiol-binding sites under *in vivo* conditions.

The method is based on the exchange of [^{3}H]estradiol with nonlabeled estradiol bound in the nuclear fraction. The nuclear fraction of estrogen responsive tissue is prepared and exposed *in vitro* at 37°C to [^{3}H]estradiol and [^{3}H]estradiol plus a 100 times excess of diethylstilbestrol (DES). The quantity of specific $R \cdot E$ complex is determined by subtracting the concentration of [^{3}H]estradiol bound in the presence of DES from the concentration of [^{3}H]estradiol bound in the absence of DES. The assay procedure has undergone rigorous validation and is completely discussed in Anderson et al. (1972a) and Clark et al. (1972, 1973b). Typical results that can be obtained using the [^{3}H]-estradiol exchange assay are shown in Fig. 2.

Tissue and hormonal specificity have also been demonstrated. Estradiol treatment results in a marked increase in nuclear $R \cdot E$ in uterus, vagina, and pituitary, while nontarget tissues (such as muscle and kidney) show little to no nuclear $R \cdot E$ following an injection of estradiol. Injection of testosterone and progesterone do not result in an accumulation of nuclear $R \cdot E$, whereas DES, estradiol, estrone, and estriol yield positive results. Nuclear $R \cdot E$ should exhibit stereospecificity for estrogenic compounds; this has been shown by adding estrogenic and nonestrogenic compounds to

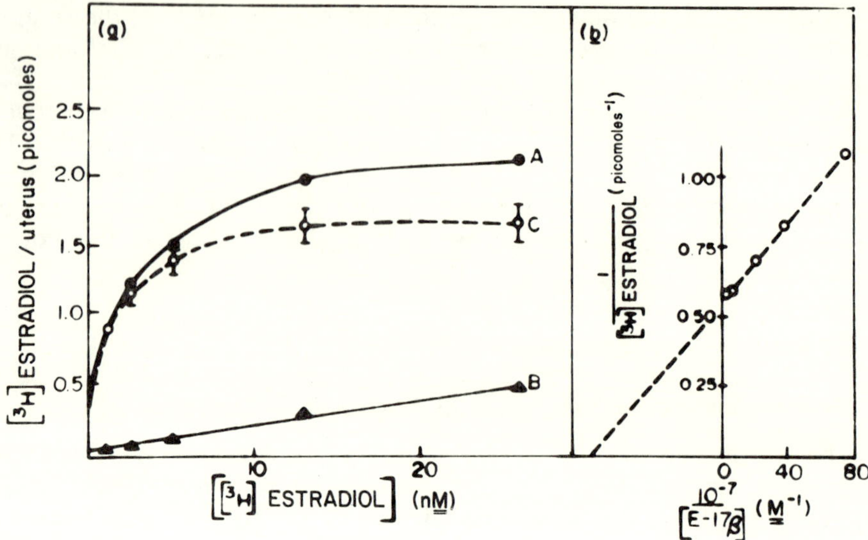

FIGURE 2. Determination of the number of specific nuclear-binding sites in the nuclear fraction of rat uteri by the [³H]estradiol exchange assay. (a) Rats were injected with 2.5 μg of estradiol. One hour after hormone administration the rats were killed and the uterine nuclear fraction prepared as described in text. This fraction was incubated for 1 hr at 37°C with either [³H]estradiol alone (A,•) or [³H]estradiol plus stilbestrol at 100 times the concentration of [³H]estradiol (B, ▲). C (○) was obtained by subtracting the amount of [³H]estradiol found with the nuclear fraction incubated with stilbestrol from the total nuclear binding in the absence of stilbestrol. Each point represents the mean of six determinations, with three rats per determination. (b) Lineweaver-Burk plot of the data represented by C (○) in part (a). The apparent K_d^R is 1.3 nM and the number of binding sites per uterus is 1.8 pmol.

the assay buffer containing [³H]estradiol. Competition for the exchange of [³H]estradiol for estradiol that was bound to the receptor was observed only for estrogenic compounds.

R·E Complex in the Nuclear Fraction of the Uterus and Pituitary during Estrus

Previous reports of the translocation and accumulation of the R·E complex in the nucleus have employed tissues that have been previously exposed to exogenous estrogen. It has been assumed that the observed translocation and accumulation of the R·E complex under these conditions also occurs *in vivo* in response to circulating endogenous estrogen. If these processes are of physiologic significance, they should occur during the estrous cycle in response to changing blood levels of estrogens. We have examined the concentration of nuclear R·E in the rat uterus and pituitary throughout this cycle. These studies were undertaken to define the relationship between the concentration of nuclear R·E complex and the hormonal state of the animal.

The concentration of nuclear R·E in the uterus was determined by the [^{3}H]estradiol exchange assay at each stage of the estrous cycle. The results are shown in Fig. 3 [also see Clark et al. (1973b)]. The number of binding sites per mg DNA is at a minimum in estrus and metestrus (approximately 1,000 sites per nucleus) (Fig. 1). The concentration increases sharply between metestrus and diestrus (approximating 3,500 sites per nucleus; $p < 0.05$) and reaches a maximum on the day of proestrus (approximately 5,000 sites per nucleus; $p < 0.05$). This relationship between the amount of R·E per nucleus and the stage of the estrous cycle is also evident when nuclear R·E is expressed on a uterine weight or protein basis.

Similar fluctuations in the quantity of nuclear R·E were observed in the pituitary during the estrous cycle. The concentrations of nuclear R·E were not significantly different at metestrus, diestrus, or estrus but showed a highly significant increase at proestrus. This increase in nuclear R·E may be an important step in the mechanism by which estrogen exerts control over gonadotropin secretion.

The cyclic fluctuation in the concentration of nuclear R·E in the uterus and pituitary during the estrous cycle closely parallels the rate of

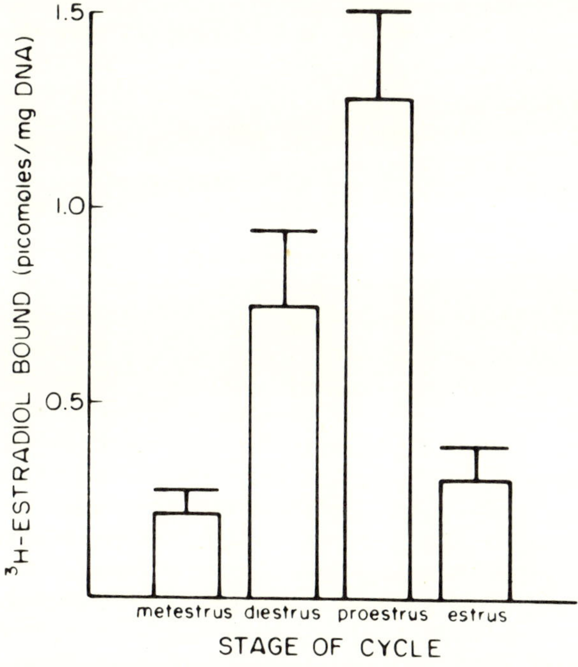

FIGURE 3. The concentration of nuclear receptor–estrogen complex in the rat uterus during the estrous cycle. The quantity of nuclear receptor–estrogen complex was determined by the [^{3}H]estradiol exchange assay for each day of the estrous cycle. Each value represents the mean ± SEM of five to six determinations.

ovarian estrogen secretion. The low concentration of nuclear R·E during estrus and metestrus is accompanied by a minimal rate of ovarian estrogen secretion. During diestrus the estrogen secretory rate increases, and this increase is accompanied by an elevation of the content of nuclear R·E in the uterus over the estrus and metestrus. In addition, the maximal concentration of nuclear R·E during proestrus coincides with the peak of estrogen secretion. This correlation suggests that the cyclic fluctuation of uterine R·E level is under the control of ovarian estrogens.

Similar K_d^R value (1.2–2.6 × 10^{-9} M, 37°C) throughout the estrous cycle and after estradiol treatment suggest that the receptors are homogenous in their affinity for estradiol. These K_d^R values are in agreement with those obtained by others for both cytoplasmic and nuclear estrogen-binding sites.

A causal relationship between the elevated uterine weight, protein level, and/or protein/DNA ratio and the maximal concentration of nuclear R·E during proestrus remains to be established. Estrogen is known to augment uterine protein and RNA synthesis—effects that are blocked by puromycin and actinomycin (Hamilton, 1968). Moreover, the R·E complex has been implicated in the stimulation of RNA polymerase activity by estrogen. While the possibility that the estrogen receptor exerts its effects at the cytoplasmic level cannot be excluded, the results of this investigation suggest that nuclear R·E complexes may be of physiologic importance and are not simply a pharmacologic phenomenon. This is the first demonstration that the translocation and accumulation of R·E actually occurs under physiologic circumstances.

Effect of Estradiol Dose on the Concentration of Nuclear R·E Complex

To evaluate the effects of both physiologic and hyperphysiologic doses of estradiol on the quantity of nuclear R·E, rats were injected with a wide range of estradiol doses (0.01–2.5 μg). The animals were killed at various times after injection, and the quantity of nuclear R·E was determined by the [^{3}H]estradiol exchange assay (Fig. 4). The concentration of nuclear R·E is at a maximum 1 hr after estradiol injection. At 1 hr, physiologic doses (0.05–0.1 μg) of the hormone do not result in the saturation of nuclear R·E. The decline in nuclear R·E between the first and sixth hours after treatment is dependent on the quantity of estradiol administered. During this period 70–80% of the nuclear R·E complexes are lost from the nuclear fraction following pharmacologic doses (1.0–2.5 μg) of the hormone, while a 7–9% decrease is observed with more physiologic doses (0.05–0.1 μg). The differential rates of disappearance of nuclear R·E during these 5 hr result in approximately equivalent quantities of nuclear R·E complex at 6 hr after estradiol injection whether nuclear R·E was produced by physiologic (0.1 μg) or pharmacologic doses (0.4–2.5 μg) of the hormone. One can speculate that physiologic doses of estradiol may

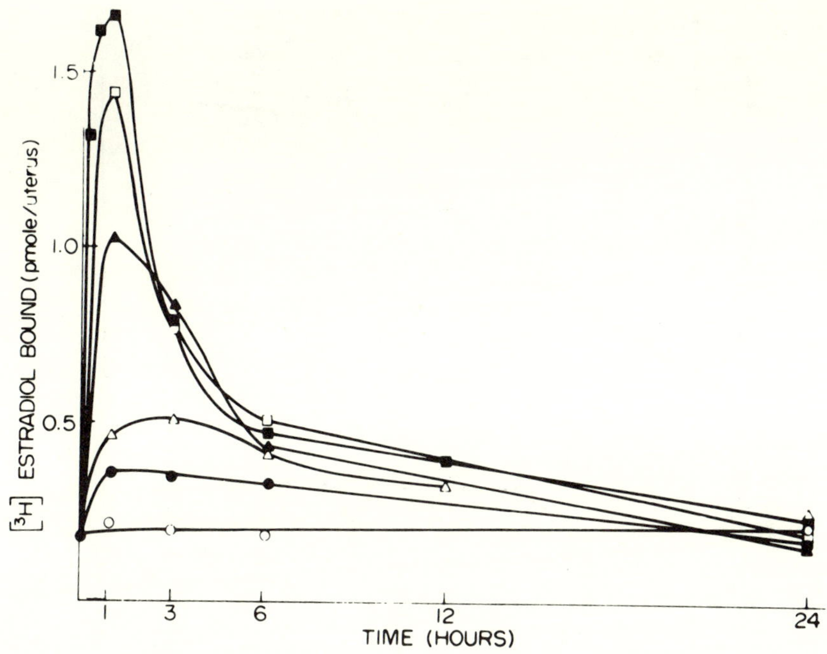

FIGURE 4. Relationship between estradiol dose and the quantity of R·E complex in the nuclear fraction of the rat uterus. Rats were injected subcutaneously with saline or estradiol: 0.01 μg (○); 0.05 μg (●); 0.1 μg (△); 1.4 μg (▲); 1.0 μg (□); or 2.5 μg (■). At the designated times after injection the animals were killed, and the quantity of nuclear R·E was determined by the [³H]estradiol exchange assay. In this figure the mean ± SEM for nuclear R·E was calculated from at least four determinations, with two rats per determination.

result in that number of nuclear receptors that can be accommodated by nuclear acceptor sites. As the hormonal dose is increased above the physiologic level, the acceptor molecules may become saturated and the excess R·E complex may be lost from the nuclear fraction. Therefore, we consider these data to be in support of the idea that the nuclei of estrogen target tissues contain a limited number of acceptor sites.

Relationship between Nuclear R·E Binding and Uterine Growth

To evaluate the relationship between early uterine responses and the stimulation of true uterine growth, several of the well-known early uterotropic responses were examined following estradiol or estriol treatment. Animals were injected with saline, estradiol, or estriol and sacrificed 3 hr following the hormone injection. The uteri were weighed and the conversion of glucose to CO_2 determined (Fig. 5). The enhancement of uterine fluid imbibition (Fig. 5A) and gluclose oxidation (Fig. 5B) elicited by the various doses of estradiol are equal to the increases induced by the same levels of estriol. Similarly, estradiol and estriol are of equal potency

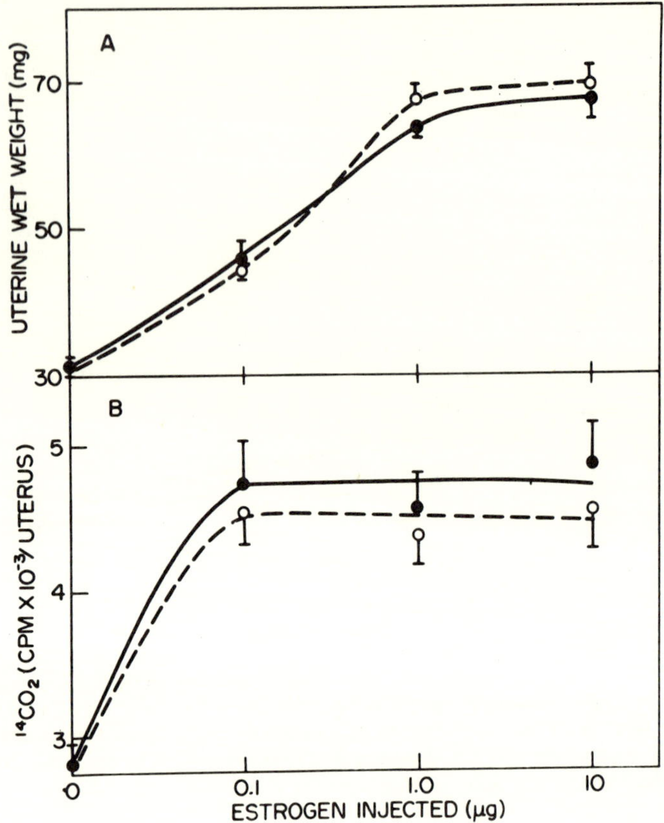

FIGURE 5. Comparison of uterine wet weight and glucose oxidation 3 hr after injection of estradiol and estriol. Rats were killed 3 hr after the injection of saline, estradiol (0.1, 1.0, and 10 µg, ● —— ●), or estriol (0.1, 1.0, and 10 µg, ○ —— ○). (A) In this and subsequent figures the mean uterine weight ± SEM was calculated from at least eight rats. (B) In this and subsequent figures the mean conversion of glucose ± SEM was calculated from at least five individual determinations.

with respect to their ability to stimulate increases in the activity of RNA polymerase and in the rates of conversion of glucose to lipid, protein, and RNA 3 hr after injection (Table 1). Thus, estradiol and estriol exhibit equal abilities to stimulate these early uterotropic responses.

If the primary function of estrogen is to induce early uterine responses that, in turn, stimulate the subsequent synthetic events responsible for growth, one would predict from the data in Fig. 5 and Table 1 that estradiol and estriol would share equal potencies with regard to the growth process. However, this is not the case (Fig. 6). The increases in uterine dry weight at 24 hr following estradiol injection are highly significant, whereas the increases evoked by estriol are only slightly above the saline control. These results show that uterine growth requires a factor(s) other than the early responses shown in Fig. 5 and Table 1, which is furnished by estradiol but not estriol.

TABLE 1. Effects of Estradiol and Estriol on Short-Term Uterine Responses[a]

Response	Saline	Estradiol	Estriol
[^{14}C] lipid			
(dpm/uterus)	2,001 ± 218	6,183 ± 604	7,272 ± 985
[^{14}C] protein			
(dpm/mg protein)	174 ± 4	320 ± 15	346 ± 25
[^{14}C] RNA			
(dpm/100 μg RNA)	2,227 ± 332	3,395 ± 319	3,570 ± 239
RNA polymerase			
(dpm/100 μg RNA)	803 ± 104	1,631 ± 219	1,850 ± 94

[a]Rats were killed at 3 hr after the injection of saline, estradiol (1 μg), or estriol (1 μg). The mean ± SEM conversion of [^{14}C] glucose to [^{14}C] lipid, [^{14}C] protein, and [^{14}C] RNA was calculated from at least four determinations with two uteri per determination. RNA polymerase activity, expressed as dpm of [^{3}H] UT incorporated into RNA/100 μg RNA, was determined by procedures described by Gorski (Anderson et al., 1973). The means ± SEM were determined from at least four determinations with three uteri per determination.

We have previously suggested that the failure of estriol to cause a significant increase in uterine weight 24 hr after treatment was related to the inability of estriol to cause long-term retention of the estrogen receptor by the nucleus (Anderson et al., 1972b). To examine this hypothesis and to describe these relationships in more detail, the amount of nuclear R·E and the extent of uterine growth were examined as a function of time following an injection of estradiol or estriol.

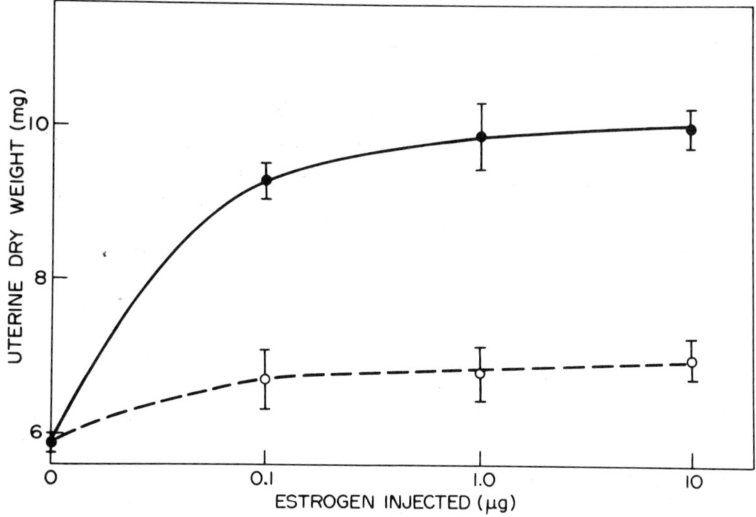

FIGURE 6. Effects of estradiol and estriol on long-term uterine growth. Rats were injected with saline, estradiol (0.1, 1.0, and 10 μg, ●——●), or estriol (0.1, 1.0, and 10 μg, ○——○). Twenty-four hours after injection the rats were killed and the uterine dry weight determined.

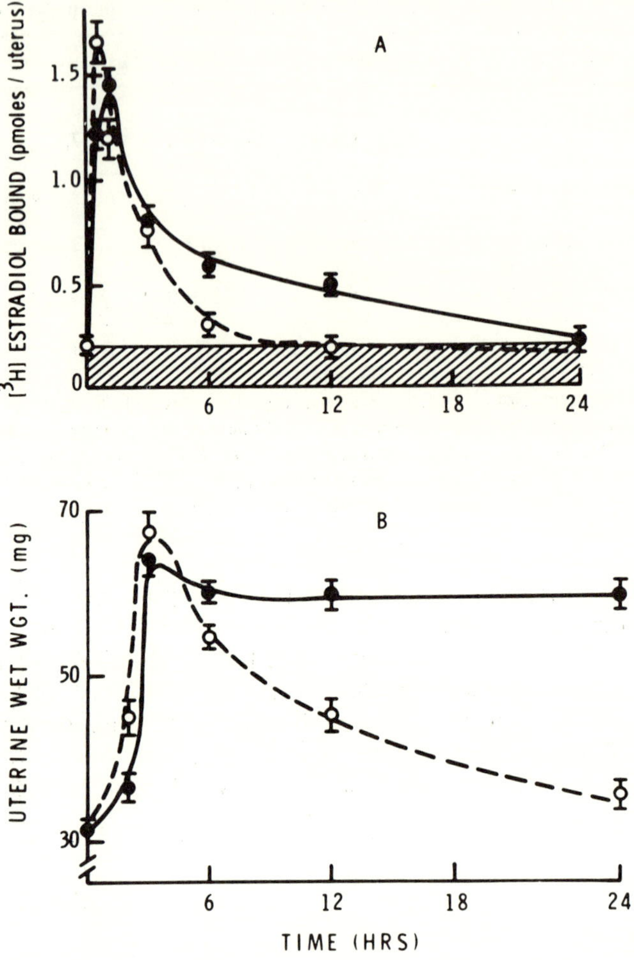

FIGURE 7. The concentration-time parameters of nuclear receptor-estrogen binding and uterine response following an injection of estradiol or estriol. Immature rats were injected with 1.0 μg of either estradiol (●) or estriol (○). At various times after treatment the uteri were weighed to the nearest 0.1 mg (B) and the concentration of the nuclear R·E complex (A) was determined by the [³H]estradiol exchange assay. The crosshatched portion in graph A represents the control lever of nuclear R·E in saline-injected controls.

Immature rats were injected with 1.0 μg of estradiol (E_2) or estriol (E_3) and the concentration of nuclear R·E was determined at various times following injection (Fig. 7A). The uterine weight was also measured at various times after injection (Fig. 7B). The concentrations of nuclear receptor-estrogen complex are equivalent at 3 hr after an injection of either estradiol or estriol (Fig. 7A). Uterine responses are also equivalent at 3 hr (Table 1, Figs. 5 and 7B). However, by 6 hr the concentration of nuclear R·E, which is elicited by estriol, has declined to near control levels, while that by estradiol remains well above control. This rapid decline in the nuclear $R \cdot E_3$ complex, compared with the $R \cdot E_2$ complex, is paralleled by a corresponding inability of estriol to maintain uterine wet weight (Fig. 7B) and to elicit long-term growth responses (Fig. 6). These experiments support the idea that the time required for the stimulation by the receptor estrogen complex of nuclear events that result in long-term uterine growth responses is at least 6 hr and that estriol is a weak estrogen in this regard because it does not promote long-term retention of the receptor within the nuclear compartment.

Estriol has long been considered a weak estrogen with regard to the stimulation of true uterine growth (Dorfman, 1940; Hisaw, 1959). The results shown in Fig. 7 suggest that the low uterotropic activity of estriol is probably due to the labile nature of the nuclear $R \cdot E_3$ complex. If this suggestion is correct, one would predict that estriol injections at 3-hr intervals, by preventing the rapid decline of the nuclear $R \cdot E_3$ complex, would lead to a marked enhancement of the uterotropic activity of the estrogen. To evaluate this possibility, rats received injections of 0.5 µg of estradiol or estriol every 3 hr for 21 hr. At 24 hr after the first injection the animals were killed, and the dry weight of the uterus was determined (Fig. 8, crosshatched bars). This injection procedure maintains similar

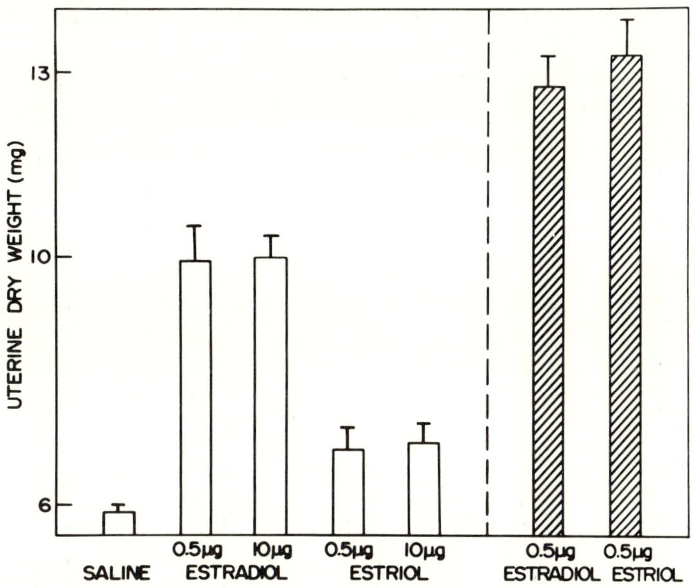

FIGURE 8. Effects of single and multiple injections of estradiol or estriol on long-term uterine growth. (Left) Rats were injected once with estradiol (0.5 and 10 µg) or estriol (0.5 and 10 µg), and the uterine weight was determined 24 hr after injection. Each value represents the mean ± SEM of three determinations. (Right) Rats were injected with estradiol (0.5 µg) or estriol (0.5 µg) every 3 hr for 21 hr (eight injections). Twenty-four hours after the first injection, the uterine dry weight was determined. Each value represents the mean ± SEM of three determinations.

concentration of the $R \cdot E_2$ or $R \cdot E_3$ complex in the nucleus for the experimental period (see below). The effects of single injections of saline, estradiol, or estriol on dry weight are also shown in Fig. 8 (open bars) for comparison. The uterine dry weight elicited by repetitive estriol injections is approximately twofold greater than the saline injected value and is not significantly different from the response induced by multiple injections of estradiol ($p < 0.5$). These results suggest that the maintenance of the $R \cdot E_2$ and $R \cdot E_3$ complex within the nucleus for 24 hr by sequential injections results in the tonic stimulation of the nuclear mechanisms responsible for uterine growth.

At least two possibilities exist to explain why the $R \cdot E_3$ complex in the nucleus of the uterine cells following a single estriol injection is not conducive to growth, whereas the presence of the $R \cdot E_3$ complex for extended time periods dramatically increases the long-term response. It may be that the $R \cdot E$ complex must be present in the nucleus at one critical time after estrogen injection in order to elicit the nuclear events required for the growth process. Alternatively, the extent of the growth response may depend on the length of time the $R \cdot E$ complex remains in the nucleus of uterine cells.

To evaluate these possibilities, rats received a variable number of estriol injections (every 3 hr up to 21 hr), and the extent of the dry weight response was assessed 24 hr after the first injection (Fig. 9). As the number of estriol

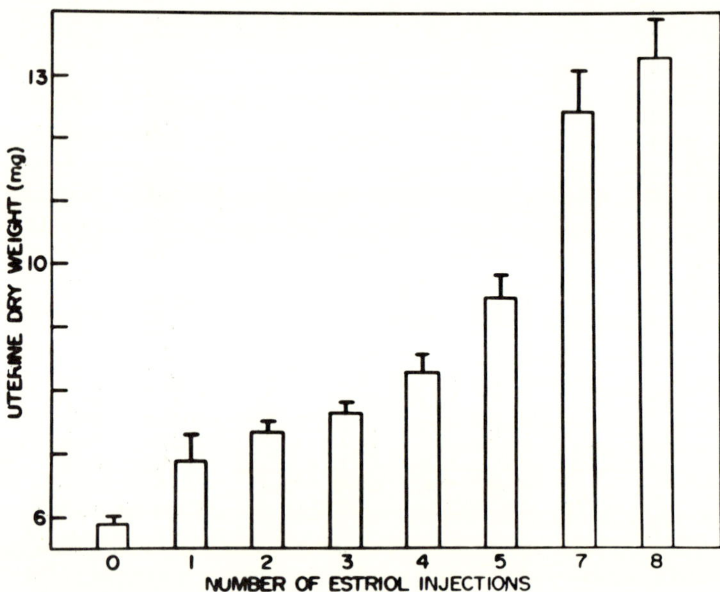

FIGURE 9. Effects of a variable number of estriol injections on long-term uterine growth. Immature rats received a variable number of injections of 0.5 μg of estriol (one to eight injections). The animals were sacrificed 24 hr after the first injection and the uterine dry weights determined.

TABLE 2. The Effect of Nafoxidine on the Quantity of Receptor in the Nucleus, Uterine Growth, and Vaginal Cytology

Treatment[a]	Nuclear receptor (pmol/uterus)	Uterine wet weight (mg)	Vaginal cytology[b]
7 days			
Saline	0.10[c]	37.0[d]	Closed
Nafoxidine	0.91	66.0	Open, L, NE
19 days			
Saline	0.23	41.2	Open, L
Nafoxidine	1.05	81.0	Open, NE, C

[a]Ovariectomized 21-day-old rats were injected with saline or 100 μg of nafoxidine, and the quantity of receptor in the uterine nuclear fraction, uterine weight, and vaginal cytology were determined 7 and 19 days after the injection.

[b]L, leukocytes; NE, nucleated epithelial cells; C, cornified cells.

[c]Each value represents the mean of 3–4 determinations with two uteri/determination.

[d]Each value represents the mean of 6–8 determinations.

injections increased, the uterus exhibited a corresponding increase in dry weight. Since the number of estriol injections can be equated with the presence of the $R \cdot E_3$ complex in the nucleus (Fig. 7A vs. Table 2), these results imply that the greater the length of time the $R \cdot E_3$ complex remains in the nucleus, the greater is the uterine growth.

Customarily, the problem of estrogen action has been reduced to the study of very early responses with the hope of identifying the initial steps thought to stimulate the subsequent synthetic growth processes. The various theories for the primary regulatory action of estrogen can be classified according to two fundamental concepts. The suggestion has been made that the early transport and metabolic responses constitute the primary stimulatory action of estrogen by providing substrates that are rate limiting for growth (Szego, 1971; Szego and Roberts, 1953). Alternatively, several authors have indicated that early changes in RNA metabolism, RNA polymerase activity, and the synthesis of specific uterine protein are primary events that result in the production of long-term uterine growth (Balieu et al., 1972; Mayol and Thayer, 1970; Talwar et al., 1973). The results of this investigation have important implications with respect to these hypotheses.

It is unlikely that the only function of the $R \cdot E$ complex is to trigger an early primary event that results in a cascade of metabolic and growth responses. Our results and the work of others have demonstrated that estradiol and estriol share similar activities with regard to the stimulation of uterine blood flow (J. N. Anderson, J. H. Clark, and E. J. Peck, Jr., unpublished), fluid imbibition (Fig. 5), glucose transport (Gorski and Baker, 1973), glucose utilization (Fig. 5; Table 1), and histamine release (Clark et al., 1974) during the first 6 hr after injection. In addition, estriol and estradiol have been reported to enhance early RNA polymerase activity (Table 1) (Clark et al., 1974a) and to stimulate the early synthesis of RNA, protein (Hamilton, 1963), and the specific estrogen-induced protein (Ruh et al., 1973). Since estriol produces marginal growth as reflected by dry weight (Fig. 6), it seems unlikely that these early responses, either alone or in combination, lead directly to the stimulation of cellular processes responsible for uterine growth. These observations suggest that the concept of an early pivotal step in the induction of uterine growth by estrogen is incorrect. While one could argue that the effects of estriol on the hypothetical primary event have not been experimentally determined, this seems unlikely since estriol and estradiol have been shown to share similar potencies regarding most of the known early uterine responses. In addition, one might expect that the known early responses would depend on the primary response; hence, their induction would be indicative of its stimulation.

The results of this investigation and a previous report from our laboratory (Anderson et al., 1972b) strongly suggest that the presence of the receptor–estrogen complex in the nucleus of uterine cells for at least 6 hr after estrogen treatment is requisite for the induction of uterine growth. First, elevated levels of nuclear R·E at 1–3 hr after injection of hyperphysiologic doses of estradiol are not retained by the nuclear fraction beyond 3 hr and do not appear to contribute to the growth responses (Anderson et al., 1973). Second, the level of nuclear R·E at 6 hr (but not at 1 or 3 hr) after estradiol injection is correlated to the long-term uterine responses measurable at 24 hr after estradiol treatment (Anderson et al., 1972b). Third, the rapid decline in the level of nuclear R·E complex between 3 and 12 hr after the administration of estriol (Fig. 7) is accompanied by a corresponding decline in the ability of estriol to induce the long-term growth response (Figs. 6 and 7). Finally, maintaining the $R·E_2$ and $R·E_3$ complexes within the nucleus for extended periods of time by repetitive hormone injections elevates the potency of estradiol and especially of estriol with respect to the induction of uterine growth (Figs. 8 and 9).

The mere stimulation of early uterotropic events does not lead to true uterine growth. It is clear that estriol is as efficacious as estradiol in the stimulation or early uterotropic events, yet this compound does not produce significant true growth as reflected by dry weight. This failure of estriol to produce true growth appears to be due to the loss of the $R·E_3$ complex from the nuclear component. Thus, the $R·E_2$ complex remains in the nucleus for a longer period of time than does the $R·E_3$ complex, and this long-term retention may be involved in the production of true uterine growth.

We have demonstrated that the translocation of the estrogen receptor to its nuclear locus takes place under the influence of endogenous estrogen during the estrous cycle. This is the first demonstration that this phenomenon occurs *in vivo* under physiologic circumstances. The fluctuating levels of nuclear R·E in the uterus and pituitary during the estrous cycle are probably involved in the reproductive control mechanisms that are influenced by estrogen.

Although the exact mechanism by which estrogens induce uterine growth is not clear, we believe that estrogen manifests its uterotropic action in two phases. Phase one consists of such well-known responses as permeability changes, release of histamine, increase in blood flow, and increase in 3′,5′-cyclic AMP. The phase one or early responses represent a period of preparation for true growth. Phase two consists of nuclear synthetic events that are controlled by the presence of nuclear R·E for approximately 6 hr or longer. Thus, the substrates and metabolic conditions provided in phase one may be employed for the production of cellular constituents characteristic of late uterotropic responses—the products of phase two.

This proposal has important implications with respect to the primary events in the mechanism of action of estrogen. It is unlikely that the only function of the R·E complex is to trigger a primary event that then results in subsequent metabolic and growth responses. Estriol has been reported to stimulate the synthesis of an estrogen-induced uterine protein, IP, and this protein has been implicated as a primary stimulatory event in inducing uterine growth. Since estriol does not produce long-term uterine growth, it seems unlikely that IP is involved in the stimulation of nuclear events that lead to increased uterine growth. Further, estriol promotes increases in fluid imbibition, glucose oxidation, and RNA and protein synthesis during the first 4 hr after injection. Thus, the long-term uterotropic responses are probably not a direct result of cascading effects initiated by these early responses.

Mechanism of Action of Estrogen Antagonist

Nonsteroidal estrogen antagonists are considered to be antiestrogenic because they partially inhibit the effects of estrogen on uterine growth and vaginal cornification (Emmens, 1970). It has been suggested that the mechanism of estrogen antagonism resides in the ability of the compounds to compete for the estrogen receptor found in the cytoplasm of estrogen target tissues and thereby to reduce the quantity of the receptor–estrogen complex formed (Jensen et al., 1966; Rochefort et al., 1972). The formation of the receptor–estradiol complex and its subsequent translocation to nuclear sites is considered to be a primary event in the mechanism of action of estrogen (Clark et al., 1973b; Jensen et al., 1973; O'Malley and Means, 1974), hence the reduction of receptor–estradiol complex would lead to a decreased estrogen stimulation.

In the work presented here we have examined the above hypothesis and will demonstrate that the mechanism by which antiestrogens antagonize estrogen-stimulated uterine growth resides in their inability to stimulate the synthesis and/or reutilization of the cytoplasmic estrogen receptor subsequent to initial nuclear binding.

Effect of Estradiol and Nafoxidine

Immature female rats (21–22 days old) were injected subcutaneously with either 2.5 µg of estradiol or 100 µg of nafoxidine dissolved in 0.5 ml of saline that contained 5% ethanol. These quantities of estradiol and nafoxidine (Upjohn 11,100 A) are sufficient to maximize uterine growth. Uterine weight was measured at various times following the injection, and the quantity of receptor in the nuclear fraction was determined by the [³H]estradiol exchange assay (Anderson et al., 1972a).

An injection of 2.5 µg of estradiol results in the rapid accumulation of receptor by the nuclear fraction that reaches a maximum by 1 hr (Fig. 10). The amount of receptor retained by the nucleus decreases rapidly between 1 and 6 hr and then more slowly until 24 hr after the injection when control values are obtained (Anderson et al., 1972b). In contrast,

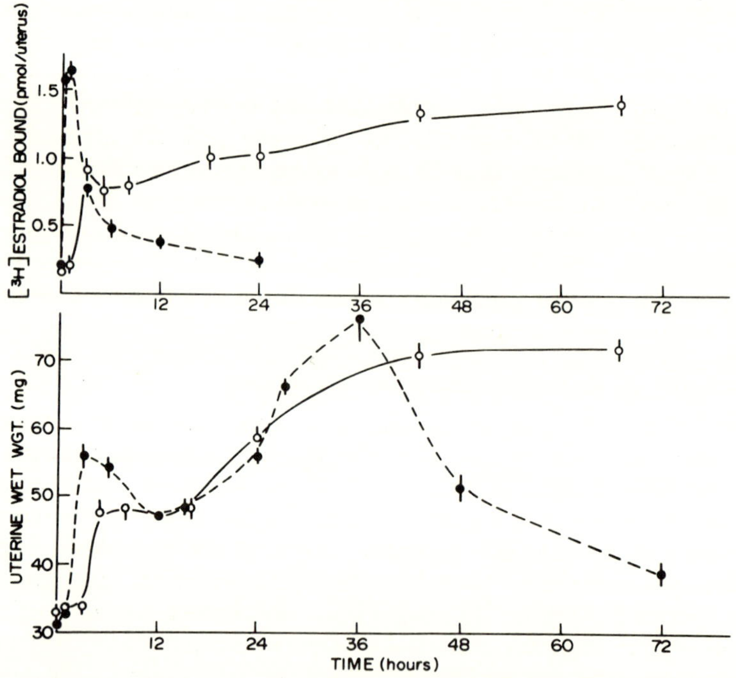

FIGURE 10. Effect of nafoxidine and estradiol on the accumulation and retention of the estrogen receptor in the uterine nuclear fraction and uterine growth. (Top) The quantity of receptor in the nuclear fraction as a result of estradiol (●) or nafoxidine (○) injection was determined by the [³H]estradiol exchange assay. The values represent the mean ± SEM. (Bottom) Uterine wet weights were measured as a function of time after an injection of estradiol (●) or nafoxidine (○). The values represent the mean ± SEM.

nafoxidine causes a slower accumulation of receptor in the nuclear fraction that essentially reaches maximum by 3 hr (Fig. 10) (Clark et al., 1973a). The quantity of receptor in the nucleus remains at this level *for at least* 67 hr. As shown in Fig. 10, uterine wet weight (as well as dry weight, not shown) reaches a maximum by 36 hr and then declines in rats treated with estradiol. The wet weight of uteri (as well as dry weights, not shown) of rats treated with nafoxidine reaches a maximum by 43 hr but does not decline. Rather, uteri are maintained in a stimulated state for at least 67 hr (Fig. 10).

To examine nafoxidine-induced retention over longer periods of time, 21-day-old rats were ovariectomized and injected with nafoxidine. The quantity of translocated receptor in the nuclear fraction and the uterine weight were measured at 7 and 19 days after injection. The quantity of receptor in the nuclear fraction at 7 and 19 days remains high and is not different from the quantity observed 3 hr after an injection (Table 2) (Clark et al., 1973a, 1974b). Similarly, the uterine weights at 7 and 19 days are elevated and similar to those observed 43 hr after treatment. The long-term estrogenic effects of nafoxidine are also demonstrated by the presence of nucleated and cornified epithelial cells in vaginal smears (Table 2).

Agonistic and Antagonistic Properties of Antiestrogens

These results clearly indicate that nafoxidine is an atypical estrogen that can stimulate uterine growth over extended periods of time (up to 19 days). Yet it is also clear from the work of others that serial injection of estrogen plus antagonists will inhibit uterine growth produced by estrogen (Dorfman, 1962; Emmens, 1970; Lerner, 1964). To examine this discrepancy, uterine weights were compared in rats that received a single injection of hormone and antagonist with those that received two injections (Fig. 11). The increase in uterine dry weight following a single injection of estradiol, nafoxidine, or estradiol plus nafoxidine is identical at 24 hr after the injection; by 48 hr nafoxidine or estradiol plus nafoxidine is superior to estradiol in the stimulation of uterine growth (Fig. 11). Therefore, following a single injection of estradiol and nafoxidine, there is no antagonism. Indeed, nafoxidine is clearly acting as an estrogen. However, as can be seen in Fig. 11, the antagonistic properties of nafoxidine can be observed when the compounds are administered at 24-hr intervals and the uterine weight determined at 48 hr. The antagonism observed after two or more injections at 24-hr intervals has been described by many investigators and constitutes a standard method for the measurement of estrogen antagonism.

Therefore, antagonism of uterine growth can only be observed following two or more injections of hormone and antagonist and not after a single injection. One possible explanation for this phenomenon is that nafoxidine stimulates cellular hypertrophy; thus, the uterine weight is

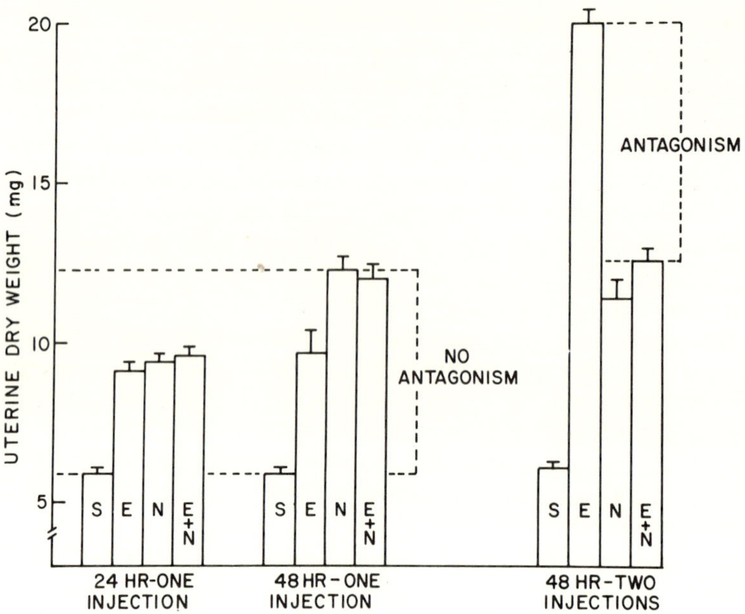

FIGURE 11. Agonistic and antagonistic effects of nafoxidine on uterine growth. Immature female rats were injected with saline (S), estradiol (2.5 μg, E), nafoxidine (100 μg, N), or estradiol plus nafoxidine (E + N) according to the following design: One group received one injection and the uterine weight was determined 24 and 48 hr after the injection. The other group received two injections 24 hr apart and uterine weight was determined 24 hr after the second injection. Each value represents the mean ± SEM for three to four determinations with five to six animals per determination.

increased, but fails to stimulate hyperplasia. Therefore, the number of cells capable of responding to subsequent estrogen injections would be reduced in nafoxidine-treated animals, which would result in a reduced uterotropic response when compared with that observed with estrogen alone.

To examine this possibility and to further characterize the growth induced by these compounds, we have measured the quantity of DNA and protein in the uterus at various times following administration of hormone and antagonist. Both estradiol and nafoxidine stimulate DNA synthesis by 48 hr. Therefore no apparent differences exist in the capacity of these compounds to elicit uterine hyperplasia (Fig. 12). The obvious difference between the uterotropic response elicited by estradiol and nafoxidine is the sustained stimulation of protein synthesis (Fig. 12). Apparently, the long-term uterotropic effects of nafoxidine arise from the ability of the receptor antagonist complex to stimulate cellular hypertrophy. From these experiments we concluded that estrogen antagonism could not be due to an inability of nafoxidine to stimulate cell growth. In fact, nafoxidine proved more effective than estradiol when uterotropic responses are compared 72 hr after treatment.

Effect of Estrogen Antagonists on Replenishing the Cytoplasmic Estrogen Receptor

Another possibility could account for the antagonism observed after two injections of hormone and antagonists: the antagonist was adversely affecting the estrogen receptor, rendering the uterus insensitive to subsequent estrogen injections. To test this hypothesis, immature rats were injected with estradiol or nafoxidine, and the quantity of cytoplasmic and nuclear receptor was determined 24 hr after treatment. The concentration of cytoplasmic receptor was determined by the [^{3}H]estradiol exchange-charcoal adsorption method as described by Clark et al. (1973a) and Katzenellenbogen et al. (1973). This method permits the evaluation of the total quantity of cytoplasmic receptor and avoids the complications that

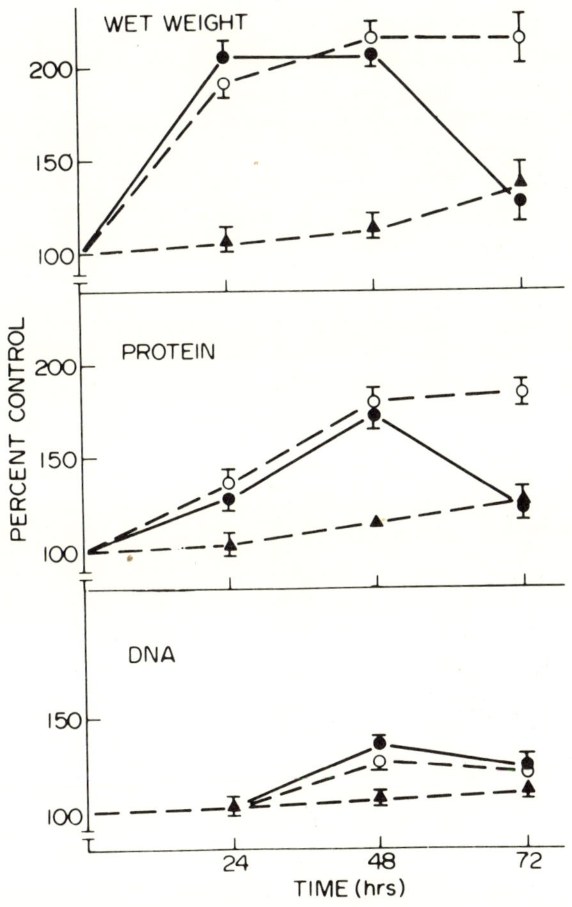

FIGURE 12. Effect of estradiol and nafoxidine on uterine protein and DNA following a single injection. Immature rats were injected with 2.5 μg estradiol (●), 100 μg of nafoxidine (○), or saline (▲). Wet weight and protein and DNA content were determined 24, 48, and 72 hr later.

arise from assays that measure only free or unfilled receptor sites (Williams and Gorski, 1971).

Estradiol treatment causes a twofold increase in the quantity of cytoplasmic receptor, compared with the saline-injected controls (Fig. 12). In contrast, nafoxidine treatment fails to stimulate the replenishment of cytoplasmic receptor; that is, the quantity of cytoplasmic receptor remains low while the amount of receptor in the nuclear fraction remains high. Thus, it appears that nafoxidine causes the translocation of cytoplasmic receptor to the nucleus where the receptor–nafoxidine complex remains bound in some unusual manner that does not result in the replenishment and/or synthesis of cytoplasmic receptor. The identity of the receptor-ligand complex in the nucleus with regard to whether the ligand is nafoxidine or a metabolite of nafoxidine has not been determined.

Regardless of the identity of the ligand, however, the accumulation and retention of the receptor by the nucleus following nafoxidine treatment is evident. The [^{3}H]estradiol exchange assay appears to measure all receptor sites that have been translocated to the nucleus under the influence of nafoxidine, since the quantity translocated is equal to the quantity of cytoplasmic sites depleted (Fig. 13). These results indicate that the uterus of the nafoxidine-treated animal would be relatively insensitive to estrogen 24 hr after treatment because the quantity of cytoplasmic estrogen receptor is very low. Thus, an injection of estrogen at this time would be less likely to result in a uterotropic response. In the

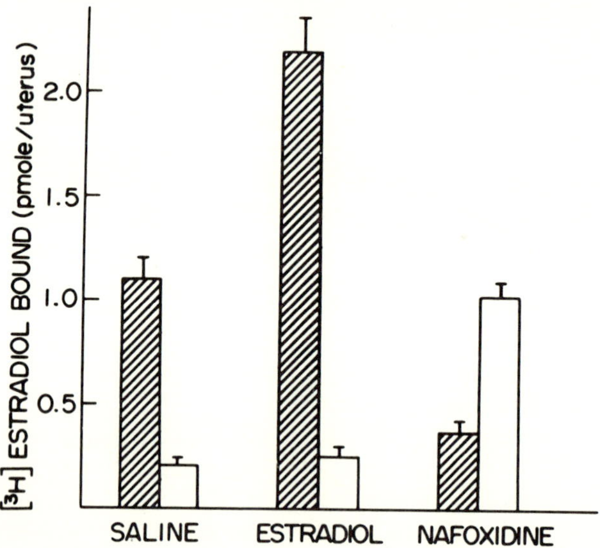

FIGURE 13. Effect of estradiol and nafoxidine on the quantity of cytoplasmic and nuclear estrogen receptor 24 hr after treatment. Immature rats were injected with saline, 2.5 μg of estradiol, or 100 μg nafoxidine. The quantity of cytoplasmic (crosshatched bars) and nuclear receptor (open bars) were determined as described in Clark et al. (1973a).

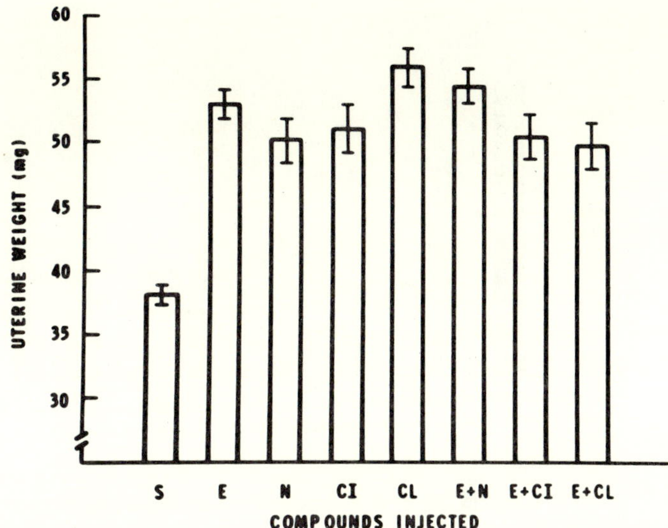

FIGURE 14. Effects of estrogen antagonists on the stimulation of uterine growth. Immature female rats were injected with either estradiol (2.5 µg, E) nafoxidine (100 µg, N), Cl-628 (0.5 mg, Cl), clomiphene (10.0 mg, CL), or a combination of estradiol plus the antagonists. The uterine weight was determined 24 hr after the injection. Each value represents the mean ± SEM for three to five determinations with four to five animals per determination.

estrogen-treated animal, however, the quantity of cytoplasmic receptor is approximately twofold that of the control, and the uterus would be very responsive to another injection of estrogen. These results explain how it is possible to observe antagonism after two injections, whereas no antagonism is observed following only one injection.

To test this hypothesis further and to establish the general nature of this phenomenon, we have examined the effects of two more estrogen antagonists, Cl-628 and clomiphene (*cis* and *trans* mixture). These compounds, along with nafoxidine, were administered alone or in combination with estradiol, and the uterine weight and the nuclear and cytoplasmic receptor content were determined 24 hr after treatment. As shown in Fig. 14, all the estrogen antagonists are clearly uterotropic and are not antagonistic to estrogen at 24 hr. Thus, Cl-628 and clomiphene are similar to nafoxidine in this regard.

The amount of receptor in the nuclear fraction is elevated significantly above the control in nafoxidine-treated rats, as was the case in Fig. 13; however, this effect is not observed with Cl-628 or clomiphene (Fig. 15). The reason for this differential effect on the quantity receptor in the nucleus is unknown and requires further study. However, all three antagonists either alone or in combination with estradiol reduce the quantity of cytoplasmic receptor by 24 hr after treatment. Thus, these antagonists exhibit the same inability to stimulate the replenishment of

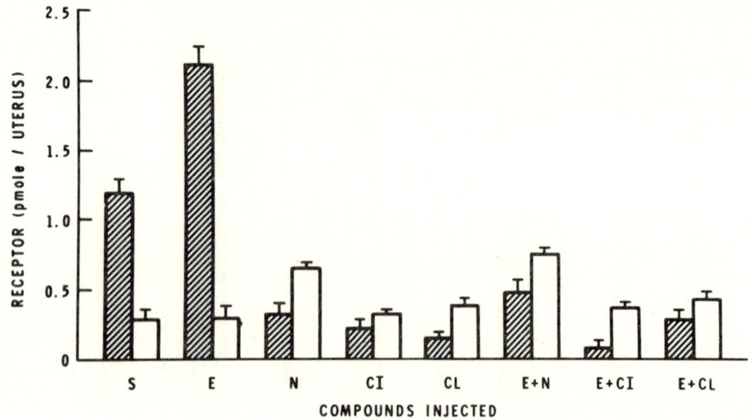

FIGURE 15. Effect of estradiol and estrogen antagonists on the uterine nuclear and cytoplasmic estrogen receptor. Immature female rats were treated with saline (S), estradiol (E), nafoxidine (N), CI-128 (CI), clomiphene (CL), or a combination of estradiol and the antagonists and the quantity of cytoplasmic (crosshatched bars) and nuclear receptor (open bars) was determined 24 hr following treatment as described by Clark et al. (1973a).

the cytoplasmic receptor, whereas estradiol treatment results in a significant increase in the number of receptors. Therefore, the ability of the uterus to respond to an injection of estradiol 24 hr following treatment with an antagonist would be greatly reduced. As we have shown with nafoxidine, subsequent estradiol treatment fails to induce any uterine growth (Fig. 11) (Clark et al., 1973a, 1974b), whereas in animals that have been pretreated with estradiol, the quantity of cytoplasmic receptor is twice that of the control and the uterus is very responsive to a second injection of estrogen (Fig. 11).

These observations explain how a compound can display agonistic and antagonistic properties and define the source of their reduced intrinsic activity. The antiestrogens are agonists because they stimulate the metabolic and regulatory pathways that cause uterine growth. The reduced intrinsic activity derives from the inability of these compounds to stimulate the replenishment of the cytoplasmic receptor; hence, they are antagonists.

It is not possible to determine from these studies whether receptor replenishment is the only uterotropic response that is not stimulated by estrogen antagonists. We feel, however, that most of the response pathways are stimulated since the increase in uterine weight induced by antagonists are equal to those induced by estradiol. These findings are important with respect to the interpretation of results obtained from experiments that use estrogen antagonist to block ovulation implantation, etc. It is possible that in these systems estrogen antagonists are stimulating estrogenic responses, just as they do in the uterus and, in a similar fashion, are not stimulating the replenishment of cytoplasmic receptor. If this were true, the involvement of the replenishment of the cytoplasmic receptor and its subsequent binding and interaction with estrogen would become an important consideration in the interpretation of results obtained in these systems.

CONCLUSION

We believe that nonsteroidal antiestrogens are antagonists not because they compete for cytoplasmic estrogen receptors and produce a complex that is a poor stimulator of growth, but because they fail to stimulate the replenishment and synthesis of the cytoplasmic estrogen receptor. This failure of estrogen antagonists to stimulate replenishment and synthesis of the cytoplasmic receptor renders the uterus insensitive to estrogen. As a result, no response greater than that seen at 24 hr is observed when the animal is subsequently injected with estrogen. In the estrogen-primed animal the uterus contains a large number of receptors and is highly responsive to estrogens.

QUESTIONS AND ANSWERS

C. Zervos: What evidence do you have to indicate that these steroids are not associated with any kind of active transport?

J. H. Clark: In a series of simple experiments, we incubated strips of diaphragm (which is similar to uterine tissue) with various concentrations of estrogen. We found that the initial entry of the fluid to this tissue is an absolute linear function of the concentration of estrogen (nonsaturable, noncompetable). Other investigators have done the same thing with individual cells at 37°C, and their data thus far indicate a simple diffusion system.

Unknown: Could not these rapid uptakes be an artifact due to the rapid uptake of water in these cells?

J. H. Clark: Estrogens stimulate many early responses. Water uptake can be blocked and the important biosynthetic events are stimulated. Therefore, water uptake could not be the cause of the subsequent stimulation of all early uterotropic events.

Unknown: Do the number of estrogen receptor sites per nucleus vary with the physiologic state?

J. H. Clark: Yes, they do vary and the number of sites are directly related to the amount of estrogen in the animal at that time. In these animals, during normal cycling, a close correspondence exists and is usually a reflection of what estrogen is doing in the blood.

Unknown: Can estrogen stimulate the synthesis of its own binding protein?

J. H. Clark: Yes, through its effect on the growth processes. The number per cell does not increase, but an increase in the number of cells results in an increase in binding and other proteins.

Unknown: Can we simply measure the number of receptors in a given tissue to obtain an indication of estrogenic activity?

J. H. Clark: No, it will not be that simple. Simply measuring cytoplasmic receptors under the best of conditions can be highly arti-factual.

W. D. Price: Will DES bind to receptor sites in the nucleus as well as the cytoplasm?

J. H. Clark: Yes, DES acts just like other estrogens.

D. A. Casciano: Have estrogen receptors been isolated and purified sufficiently to produce antibodies?

J. H. Clark: Yes, Roy Smith in our department has recently isolated and purified the estrogen receptors from chicken oviduct.

D. L. Greenman: Does E_3 cause replenishment of receptor?

J. H. Clark: Yes, in fact I think that accounts somewhat for its antiestrogenicity. It does not replenish the cytoplasmic receptor to the same degree.

J. J. Oleson: Have you also tried this work in other estrogen-sensitive tissues?

J. H. Clark: Yes, we have tried the pituitary, the hypothalamus, the mammary gland, human endometrium, DMBA-induced cancers, and mammary tissue from the gland.

REFERENCES

Anderson, J., Clark, J. H. and Peck, E. J., Jr. 1972a. Oestrogen and nuclear binding sites: Determination of specific sites by [^{3}H]oestradiol exchange. *Biochem. J.* 126:561–567.

Anderson, J. N., Clark, J. H. and Peck, E. J., Jr. 1972b. The relationship between nuclear receptor estrogen binding and uterotrophic responses. *Biochem. Biophys. Res. Commun.* 48:1460–1468.

Anderson, J. N., Clark, J. H. and Peck, E. J., Jr. 1973. Nuclear receptor estrogen complex: Accumulation retention and localization in the hypothalamus and pituitary. *Endocrinology* 92:1488.

Balieu, E. E., Wira, C. R., Milgrom, E. and Raynaud-Jammet, C. 1972. Estrogen binding and the control of RNA and protein synthesis. In *Karolinska symposia of research methods in reproductive endocrinology: Gene transcription in reproductive tissues,* ed. E. Dicsfalusy, p. 396. Stockholm: Karolinska Institutet.

Clark, J. H., Anderson, J. N. and Peck, E. J., Jr. 1972. Receptor-estrogen complex in the nuclear fraction of rat uterine cells during the estrous cycle. *Science* 176:528–530.

Clark, J. H., Anderson, J. N. and Peck, E. J., Jr. 1973a. Estrogen receptor anti-estrogen complex: A typical binding by uterine nuclei and effects on uterine growth. *Steroids* 22:707–718.

Clark, J. H., Anderson, J. N. and Peck, E. J., Jr. 1973b. Nuclear receptor estrogen complexes of rat uteri: Concentration-time-response parameters. In *Receptors for reproductive hormones,* ed. B. W. O'Malley and A. R. Means, vol. 36, pp. 15–59. New York: Plenum Press.

Clark, J. H., Glasser, S. R. and Peck, E. J., Jr. 1974a. The mechanism of action of non-steroidal estrogen antagonists. Program of the 56th annual meeting of the Endocrine Society, p. 87, Abstr.

Clark, J. H., Anderson, J. N. and Peck, E. J., Jr. 1974b. Oestrogen receptors and antagonism of steroid hormone action. *Nature* 251:446.

Dorfman, R. I. 1940. Comparison of histamine release by estradiol and estriol. *Proc. Soc. Exp. Biol. Med.* 45:594.

Dorfman, R. I. 1962. Anti-estrogenic compounds. In *Methods in hormone research*, ed. R. I. Dorfman, vol. 2, pp. 113–126. New York: Academic Press.

Emmens, C. W. 1970. Antifertility agents. *Annu. Rev. Pharmacol.* 4:237–254.

Gorski, J. and Baker, B. 1973. Early and late effects of estradiol-17B and estriol on 2-deoxyglucose metabolism in the rat uterus. Program of the 55th annual meeting of the Endocrine Society, p. 66, Abstr.

Gurpide, E. and Welch, M. 1969. Evaluation of estrogen uptake by endometrial perfusion. *J. Biol. Chem.* 244:5159–5169.

Hamilton, T. H. 1963. Isotopic studies on estrogen-induced accelerations of ribonucleic acid and protein synthesis. *Proc. Natl. Acad. Sci. U.S.A.* 49:373.

Hamilton, T. H. 1968. Control by estrogen of genetic transcription and translation. *Science* 161:649.

Hisaw, F. L., Jr. 1959. Comparative effectiveness of estrogens on fluid imbibition and growth of the rat's uterus. *Endocrinology* 64:276.

Jensen, E. V. and DeSombre, E. R. 1972. Mechanism of action of the female sex hormones. *Annu. Rev. Biochem.* 41:203–230.

Jensen, E. V., Jacobson, H. I., Flesher, J. W., Saha, N. N., Grupta, G., Smith, S., Colucci, V., Shiplacoff, D., Neumann, H. G. and DeSombre, E. R. 1966. Estrogen receptors in target tissues. In *Steroid dynamics*, ed. G. Pincus, T. Nakao and J. F. Tait, pp. 133–157. New York: Academic Press.

Jensen, E. V., Suzuki, T., Kawashima, T., Stumpf, W. E., Jungblut, P. and DeSombre, E. R. 1968. A two-step mechanism for the interaction of estradiol with rat uterus. *Proc. Natl. Acad. Sci. U.S.A.* 59:632–638.

Jensen, E. V., Mohla, S., Brecher, P. I. and DeSombre, E. R. 1973. Estrogen receptor transformation and nuclear RNA synthesis. In *Receptors for reproductive hormones*, ed. B. W. O'Malley and A. R. Means, vol. 36, pp. 60–79. New York: Plenum Press.

Katzenellenbogen, J. A., Johnson, H. J., Jr. and Carlson, K. E. 1973. Studies on the uterine, cytoplasmic estrogen binding protein. Thermal stability and ligand dissociation rate. An assay of empty and filled sites by exchange. *Biochemistry* 12:4091–4099.

Lerner, L. J. 1964. Hormonal antagonists: Inhibitors of specific activities of estrogen and androgen. *Recent Progr. Hormone Res.* 20:119–123.

Mayol, R. F. and Thayer, S. A. 1970. Synthesis of estrogen-specific proteins in the uterus of the immature rat. *Biochemistry* 9:2484.

O'Malley, B. W. and Means, A. R. 1974. Female steroid hormones and target cell nuclei. *Science* 183:610.

Peck, E. J., Jr., Burgner, J. and Clark, J. H. 1973. Estrophilix binding sites of the uterus. Relation to uptake and retention of estradiol *in vitro*. *Biochemistry* 12:4596.

Rochefort, H., Lignon, F. and Capony, F. 1972. Formation of estrogen nuclear receptor in uterus: Effect of androgens, estrone and nafoxidine. *Gynecol. Invest.* 3:43–62.

Ruh, T. S., Katzenellenbogen, B. S., Katzenellenbogen, J. A. and Gorski, J. 1973. Estrone interaction with the rat uterus: In vitro response and nuclear uptake. *Endocrinology* 92:125.

Shelesnyak, M. C. 1959. Effect of estrogens on histamine release in the rat uterus. *Proc. Soc. Exp. Biol. Med.* 100:739.

Shyamala, G. and Gorski, J. 1969. Estrogen receptors in the rat uterus. Studies on the interaction of cytosol and nuclear binding sites. *J. Biol. Chem.* 244:1097–1103.

Szego, C. M. 1971. The lysosomal membrane complex as a proximate target for steroid hormone action. In *The sex steroids*, ed. K. W. McKerns, p. 1. New York: Appleton-Century-Crofts.

Szego, C. M. and Roberts, S. 1953. Steroid action and interaction in uterine metabolism. *Recent Progr. Hormone Res.* 8:419.

Talwar, G. P., Jailkhani, B. L., Narayahan, P. R. and Narasimhan, C. 1973. Oestrogen induced synthesis of a protein in avian liver. In *Karolinska symposia on research methods in reproductive endocrinology: Protein synthesis in reproductive tissue*, ed. E. Dicsfalusy, p. 341. Stockholm: Karolinska Institutet.

Williams, D. and Gorski, J. 1971. A new assessment of subcellular distribution of bound estrogen in the uterus. *Biochem. Biophys. Res. Commun.* 45:258–264.

Received June 4, 1975
Accepted June 27, 1975

STUDIES ON THE BIOLOGICAL DISPOSITION OF DIETHYLSTILBESTROL IN RATS AND HUMANS

L. J. Fischer, J. L. Weissinger, D. E. Rickert, K. L. Hintze

The Toxicology Center, Department of Pharmacology,
University of Iowa, Iowa City, Iowa

Studies were conducted to elucidate the processes involved in the biological disposition of diethylstilbestrol (DES). The apparent biological half-life of [^{14}C]DES decreased with increasing age in immature rats. The half-life of DES in adult rats (14 min) was reached between the 15th and 25th day after birth.

Comparison of the intestinal absorption of [^{14}C]DES and [^{14}C]DES monoglucuronide showed the unconjugated drug to be absorbed much more rapidly. Hydrolysis of the conjugate took place in the lower intestine, and this permitted the drug to undergo enterohepatic cycling.

Placental transfer of [^{14}C]DES was studied in pregnant rats at mid and late gestation. The drug and its metabolites were restricted in their movement from mother to fetus, and relatively low fetal levels were found. Only 0.0006% of the maternal dose of DES was found as the unchanged drug in each fetus at 90 min after drug administration. Conjugated metabolites were found in the fetus just prior to birth but not in the 13-day-old fetus.

Analysis of plasma and urine after oral administration of both [^{3}H]DES and [^{14}C]DES monoglucuronide to two human subjects showed that the glucuronide was absorbed more slowly than DES. Similar metabolic products were found in the urine after administration of DES or its glucuronide conjugate. This and other indirect evidence indicated that the conjugate was hydrolyzed to DES in the gastrointestinal tract prior to absorption. The data showed that the biological disposition of DES in humans was not extensively different from that observed in laboratory animals.

INTRODUCTION

The potent synthetic estrogen diethylstilbestrol (DES) is widely used in clinical medicine for estrogen replacement or treatment and in higher doses as an abortifacient. Use of DES as a growth stimulant in cattle and other ruminants has allowed the possible ingestion of the estrogen and its metabolites as residues in meat. Since little is known of the biological disposition of DES, we have continued to conduct experiments on the kinetic aspects of the metabolism and excretion of this estrogen.

This research was supported by USPHS grant no. GM 12675.

J. L. Weissinger's present address is Department of Pharmacology, University of New Mexico, Albuquerque, New Mexico. D. E. Rickert's present address is Department of Pharmacology, Michigan State University, East Lansing, Michigan.

Requests for reprints should be sent to L. J. Fischer, The Toxicology Center, Department of Pharmacology, University of Iowa, Iowa City, Iowa 52242.

Journal of Toxicology and Environmental Health, 1:587–605, 1976

This report summarizes the results obtained in laboratory animals and man concerning the enterohepatic circulation and excretion of DES. Some of the results have been published and a brief account of those studies will be presented. A more detailed account will be given of previously unpublished results on the distribution and placental transfer of DES in pregnant rats and on the excretion of metabolites after oral ingestion of DES and DES monoglucuronide (DESG) by humans. The results presented here should aid in the design of future experiments concerning the assessment of possibly harmful effects caused by the ingestion of DES.

RESULTS AND DISCUSSION

Age-Related Changes in the Disposition of DES in Rats

Conjugation to glucuronic acid is a major pathway in the metabolism of DES in rats. These drug metabolites are excreted primarily in the bile (Fischer et al., 1966; Hanahan et al., 1953; Klaassen, 1973). Enterohepatic circulation of biliary metabolites of DES was demonstrated and evidence was obtained to implicate in this process the hydrolysis of conjugates by bacterial β-glucuronidase (Fischer et al., 1966).

Newborn animals are known to be deficient in glucuronide conjugation (Dutton, 1966) and would lack bacterial β-glucuronidase in the intestinal lumen at birth. It was hypothesized, therefore, that the enterohepatic circulation of DES and metabolites would be deficient in newborn rats and that development of this pathway would occur after birth. To examine this possibility, we determined the amounts of unconjugated material in the whole body of immature rats given a 17 μg/kg dose of [^{14}C]DES (Fischer and Weissinger, 1972). Animals were studied on the day of birth and at 5, 15, 25, and 50 days after birth. Figure 1 shows the results of those studies.

Rats studied on the day of birth showed a first-order decay in unconjugated radioactive material over the entire experimental period with a half-life of 131 min. Five-day-old animals had a biphasic disappearance with an initial phase half-life of 55 min. In 15- and 25-day-old animals the linear disappearance phase was shorter and half-lives of 33 and 14 min, respectively, were observed. In the older animals the unconjugated radioactive material in the body increased during the second phase. Experiments in 50-day-old rats gave a pattern of drug disappearance identical to 25-day-old animals. This indicated that drug disposition pathways for DES in rats reached maturity between 15 and 25 days after birth.

Further experiments showed that the second phase of the disappearance curves shown in Fig. 1 represented intestinal hydrolysis of biliary excreted glucuronide conjugates of DES and its hydroxylated metabolites. The results indicated that this deconjugation mechanism reached maturity at the same time as DES conjugation, i.e., between 15 and 25 days after birth.

Some evidence was obtained in these experiments to indicate that DES

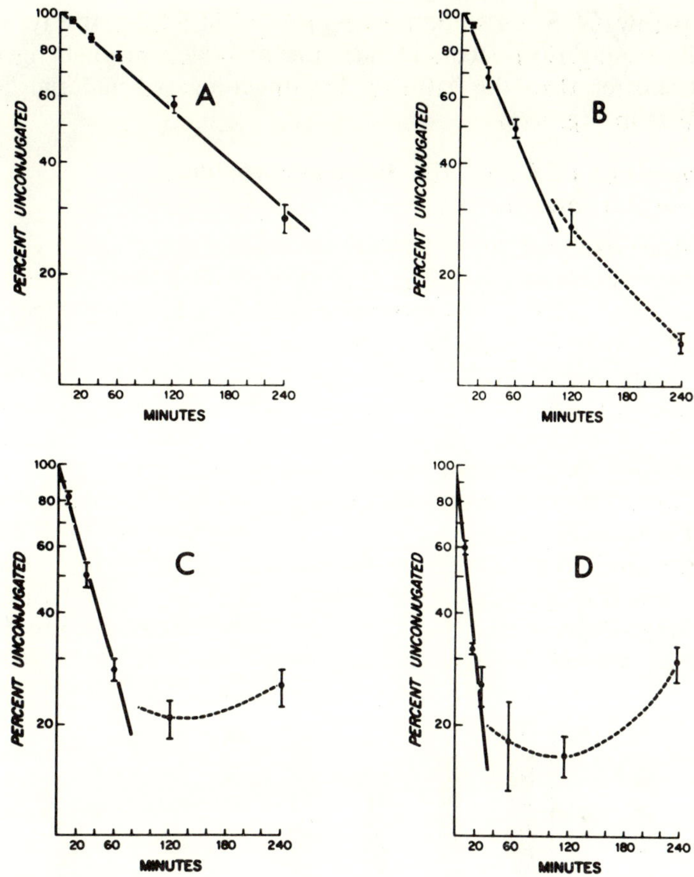

FIGURE 1. Curves showing the disappearance of unconjugated material in the whole body of immature rats given 17 μg/kg dose of [^{14}C]DES. (A) rats studied on the day of birth (0 days old); (B) 5 days old; (C) 15 days old; (D) 25 days old. The animals were sacrificed at various times after the dose and the radioactive materials in the whole body extracted prior to analysis for unconjugated and conjugated DES-related materials. Solid lines represent a regression analysis on the linear or conjugation phase of the disappearance curves. Broken lines represent the phase characterized by intestinal deconjugation. Each data point represents the mean ± SE for 6–12 animals. From Fischer and Weissinger (1972).

was metabolized to products other than DES glucuronide (Fischer and Weissinger, 1972). Reverse isotope dilution analysis for DES remaining in the body of 25-day-old rats at 4 hr after a 17 μg/kg dose of ^{14}C-labeled drug indicated that only 38% of the unconjugated radioactive material was unchanged DES. At the same time in 0-day-old animals 88% was unchanged DES. These results suggest that nonconjugative metabolism of DES is low at birth and develops with increasing age. An exhaustive study of the development of the nonconjugative metabolism of DES has not been made, and it is not known when immature rats reach the adult level for metabolism via those pathways. Since some metabolism of DES by pathways other than glucuronide conjugation occurs in rats, the half-lives for unchanged DES are somewhat shorter than the half-lives for unconjugated material determined from the data in Fig. 1.

Absorption of DES and DES Monoglucuronide from the Rat Intestine

To further examine the intestinal deconjugation phase involved in the enterohepatic circulation of DES, experiments were conducted to compare the absorption of DES and its monoglucuronide conjugate (DESG) from the intestine of 25-day-old rats (Fischer et al., 1973). Either [^{14}C]DESG or [^{14}C]DES was injected into a 7–8 cm closed loop of intestine in rats anesthetized with pentobarbital. At various times after placing the drug or its conjugate into the lumen, the intestinal loop was excised, and the amount of remaining radioactive material was separated into unconjugated and conjugated fractions. The results obtained using loops of the proximal portion of the small intestine are shown in Fig. 2. The data show that DES disappeared rapidly from the intestine when compared with DESG, which was not well absorbed from this section of the intestine. Two other facts are apparent from the data: DESG is not hydrolyzed to unconjugated DES in this part of the intestine; a small amount of conjugated material is in the intestinal

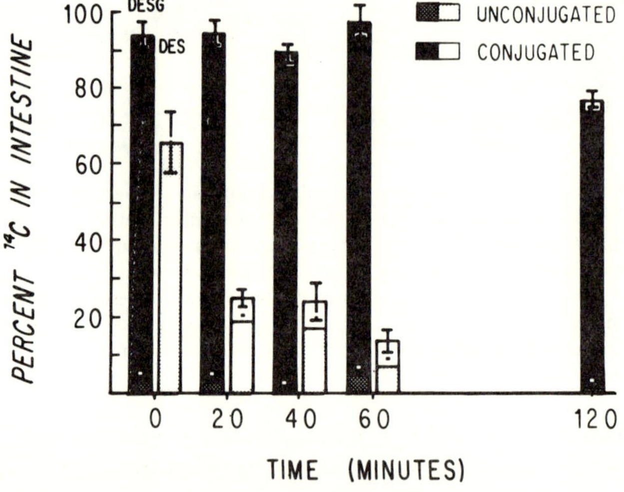

FIGURE 2. Absorption of DESG and DES from the proximal small intestine of the 25-day-old rat. The first bar at each time point represents the results of experiments in which [^{14}C]DESG was placed into the lumen of closed intestinal loops. The second bar at each time point depicts the results of identical experiments with [^{14}C]DES. Bar heights represent the total radioactivity remaining in the intestinal contents (mean ± SE, $N = 3$). The total radioactivity was separated into unconjugated material, represented by the stippled area in each bar (mean ± SE), and conjugated material, shown as the solid area. From Fischer et al. (1973). ©1973 The Williams & Wilkins Co., Baltimore.

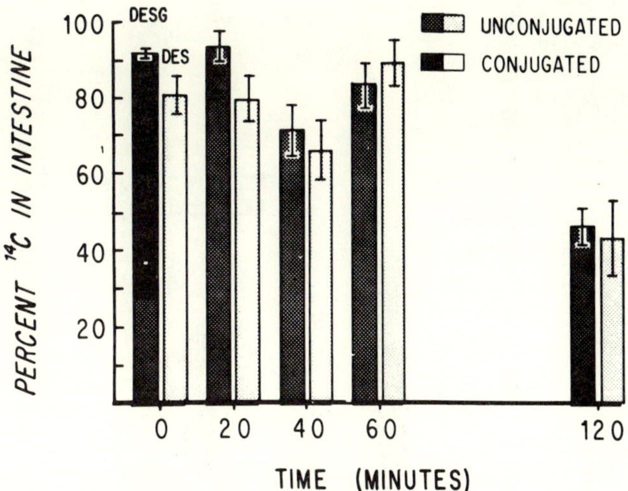

FIGURE 3. Absorption of DESG and DES from the cecum of the 25-day-old rat. Key same as Fig. 2. From Fischer et al. (1973). ©1973 The Williams & Wilkins Co., Baltimore.

contents after DES is placed in the loop. This conjugated material presumably arises from glucuronidation of DES in the intestinal wall during absorption and subsequent movement of the conjugate into the lumen. Previous *in vitro* studies (Fischer and Millburn, 1970) using everted sacs of rat intestine were consistent with the results obtained using *in vivo* loops of intestine. The everted intestinal sac experiments showed that DESG entered intestinal tissue with difficulty compared with DES and that the unconjugated drug could be conjugated with glucuronic acid during absorption.

As shown in Fig. 3, there were no differences in the rate of disappearance of radioactive material after placing [^{14}C] DES or its labeled conjugate into the cecum. This lack of difference is not surprising since the conjugate was rapidly hydrolyzed to DES in the cecal contents. It was also apparent that absorption of DES was much slower from the cecum than from the proximal small intestine. Further experiments showed that the presence of fecal material retards the absorption of DES from the rat intestine (Fischer et al., 1973). The slow absorption of DES from the cecum can be attributed to two factors: (1) the lower mucosal surface area in that area of the intestine and (2) the binding of drug to cecal contents.

The absorption experiments showed that the glucuronic acid conjugate of DES, after entering the intestine via the bile, would not be absorbed as such from the intestine. The conjugate would move down the intestinal tract to the distal small intestine and cecum where it would be hydrolyzed to DES and then slowly absorbed.

Similar absorption studies were performed using 5-day-old rats (Fischer et al., 1973). Figure 4 shows the results of those experiments. Analysis of the material remaining in the intestine at 60 min indicated that DESG was

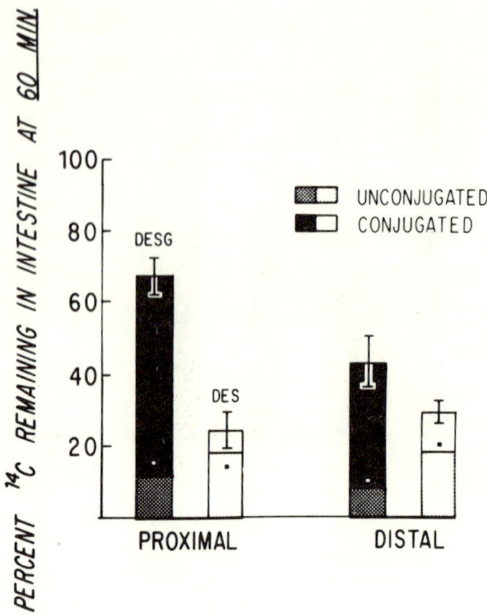

FIGURE 4. Absorption of DESG and DES from the proximal and distal intestine of 5-day-old rats. Bar heights represent total radioactivity remaining in the intestinal tissue and contents at 60 min (mean ± SE, $n = 6$). Key same as in Fig. 2. From Fischer et al. (1973). ©1973 The Williams & Wilkins Co., Baltimore.

not rapidly hydrolyzed in the gastrointestinal tract of these young animals. In spite of this, the conjugate appeared to be absorbed from the intestine, particularly from the distal portion, at a rate similar to unconjugated DES. These results indicate that the 5-day-old intestine may be much more permeable to conjugated DES than the adult intestine. Enterohepatic cycling, therefore, might occur to some extent in newborn rats without prior hydrolysis of conjugates.

Alteration of the Uterotropic Action of DES by Phenobarbital Pretreatment

Phenobarbital, a known inducer of hepatic microsomal drug metabolism, was shown by Levin et al. (1968) to influence DES metabolism and uterotropic action. Using a nonconjugating rat hepatic microsomal system, it was shown that DES was metabolized faster *in vitro* as a result of pretreatment of animals with phenobarbital. In addition, the ability of DES, given orally, to cause an increase in the uterine weight of immature rats was decreased by phenobarbital pretreatment. We extended the experiments of Levin et al. by examining the influence of phenobarbital on DES conjugation *in vivo* (Fischer and Weissinger, 1972). Pretreatment with the barbiturate (37 mg/kg/day for 4 days) was found not to alter the half-life of unconjugated [¹⁴C]DES-related material in 25-day-old rats after a single ip dose (10 μg/kg) of the labeled estrogen. It was concluded that the conjugation of DES was not altered by phenobarbital pretreatment and that if the barbiturate had an effect, it was on the nonconjugative pathways for DES metabolism.

Although Levin et al. (1968) showed that phenobarbital decreased the uterotropic action of an oral dose of DES, we were unable to reproduce this phenobarbital effect after an ip dose of the estrogen. Experiments were performed to examine possible routes of administration differences in the interaction of the metabolic enzyme inducer and DES. Groups of immature rats (phenobarbital pretreated and controls) were given a 10 μg/kg dose of [^{3}H] DES by different routes of administration, and the uterine weight and total radioactivity in the uterus were determined at 4 hr. The results are presented in Table 1. No significant differences between control and phenobarbital-treated animals were observed when the estrogen was given by the iv and ip routes. However, differences between barbiturate-treated and control animals were apparent after an oral dose of DES. Oral administration of [^{3}H] DES produced lower amounts of labeled material in the uterus at 4 hr and uterine weights were significantly lower than after estrogen administration by parenteral routes. Undoubtedly, the lower uterotropic action of DES after oral administration was due to less drug reaching the target tissue.

There could be several reasons why phenobarbital pretreatment reduced the effect of DES only when estrogen was given by the oral route. It is known that DES is metabolized to some extent during absorption from the intestine (Fischer and Millburn, 1970), and phenobarbital may have stimulated intestinal metabolism. Alternatively, a somewhat slower

TABLE 1. Effect of Phenobarbital Pretreatment and Route of Administration on the Uterotropic Action and Uterine Content of [^{3}H] DES in Immature Rats[a]

Route	N	Control	$p < 0.05$	Phenobarbital
		Uterine wet weight (mg) at 4 hr after dose		
Intravenous	7	47.2 ± 5.5	—	38.1 ± 3.2[b]
Intraperitoneal	6	38.1 ± 3.3	—	34.8 ± 1.9
Oral	8	30.0 ± 1.9	+	23.2 ± 1.1
		Picogram equivalents of [^{3}H] DES in uterus at 4 hr		
Intravenous	7	283 ± 25	—	224 ± 20
Intraperitoneal	6	227 ± 15	—	245 ± 31
Oral	8	109 ± 7	+	62 ± 5

[a]Twenty-day-old rats were treated with phenobarbital (35 mg/kg/day for 4 days) intraperitoneally and the control group received saline vehicle. Twenty-four hours after the last phenobarbital dose, the animals were given 10 μg/kg [^{3}H]DES. Four hours after the estrogen dose the animals were sacrificed and the uterus excised and weighed. The uterine tissue was subjected to combustion analysis for determination of total radioactivity in the sample.
[b]Values are the mean ± SE. Data were subjected to an analysis of variance.

absorption of the estrogen from the GI tract after an oral dose may have allowed the induced hepatic enzymes to clear more of the drug from the portal blood during its first pass through the liver. Although the mechanism is uncertain, it is clear that the ability of an inducer of drug metabolism to modify the estrogenic action of DES depends on the route of DES administration.

Maternal–Fetal Transfer of DES in the Rat

There is evidence that administration of DES to pregnant women is associated with an increased risk of vaginal carcinoma in the offspring (Greenwald et al., 1971; Herbst et al., 1971). If DES were to cause a direct alteration in fetal vaginal cells, it must cross to the fetus from the maternal circulation. Unfortunately, very little is known of the quantitative aspects of the maternal–fetal transfer of DES. In the only report available, whole-body autoradiography using mice showed that radioactivity associated with maternally administered [^{14}C]DES reached the fetus and localized in the same tissues in which high concentrations were observed in the mother (Bengtsson and Ullberg, 1963).

It is generally accepted that the maternal–fetal barrier, like the blood–brain barrier, is more permeable to lipid-soluble compounds than to their more water-soluble metabolites (Goldstein et al., 1968). Supporting this concept are the results of placental transfer studies with estrogens and their glucuronic acid conjugates. These studies showed that unconjugated estrogens passed readily to the fetus in contrast to their polar conjugates that moved to and from the fetus with difficulty (Diczfalusy et al., 1963; Goebelsmann et al., 1966; Levitz, 1966). Since conjugation with glucuronic acid is a major pathway in the rat for the biotransformation of DES, we followed the concentrations of conjugated and unconjugated [^{14}C]DES-associated radioactivity in the fetus and selected maternal tissues at various times after an intravenous dose of [^{14}C]DES. Rats were studied on days 13 and 20 of pregnancy. The experiments were designed to examine the temporal aspects of the maternal–fetal transfer of DES and to determine whether the production of polar conjugates of this synthetic estrogen alters its transplacental distribution in a manner similar to the natural estrogens.

A 0.2 mg/kg (2 μCi) dose of [^{14}C]DES (Amersham-Searle, 27.3 mCi/mmol) was injected into the tail vein of pregnant Sprague-Dawley rats in 0.5 ml of a 10% aqueous solution of propylene glycol. The animals were sacrificed by decapitation and total radioactivity in selected tissues determined by combustion analysis. Portions of the tissues were also homogenized in 20 volumes of methanol (40 volumes for liver tissue) and an aliquot of the methanol extract taken for analysis of water-soluble (conjugated) and lipid-soluble (unconjugated) [^{14}C]DES-associated radioactivity according to the method of Fischer and Weissinger (1972). The recovery of tissue radioactivity in the methanol extracts was 81–94%.

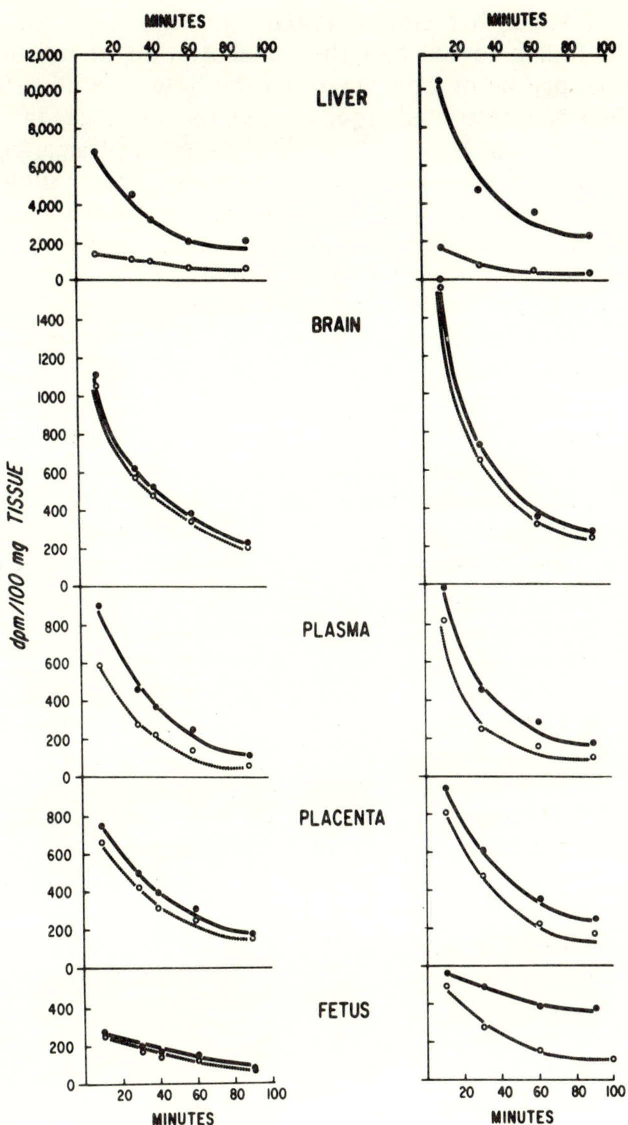

FIGURE 5. Total (—•—) and unconjugated (--○--) radioactive materials in tissues following a 0.2 mg/kg intravenous dose of [^{14}C]DES to rats on day 13 (left panel) or day 20 (right panel) of pregnancy. Values shown are the mean of three animals. Standard error of the mean values is <20%.

Comparisons of total and unconjugated radioactive materials in various tissues on days 13 and 20 of pregnancy are made in Fig. 5. Differences in the distribution and decline of unconjugated ^{14}C in maternal tissues in middle and late gestation were not observed. The rate of decline of drug-associated radioactivity was similar in all tissues studied except for the fetus where a somewhat slower initial rate was observed. Fetal tissue contained one-fourth to one-half the concentration of radioactivity found in the maternal plasma or placenta at 10 min after the dose. This indicates that [^{14}C] DES was restricted, probably in the placenta, in its movement to the fetus. The 13-day-old fetus, unlike maternal plasma, contained only unconjugated [^{14}C] DES-associated radioactivity. This shows that DES conjugates, like those of the natural estrogens, do not readily cross the placental barrier. Conjugated radioactive material, however, was found in the 20-day-old fetus. The increasing percentage of total radioactivity found as conjugated material in the fetus over the experimental period suggests that the 20-day-old fetus had developed the ability to conjugate DES. We have previously reported that the 21-day-old rat fetus is capable of conjugating DES (Fischer and Weissinger, 1972). It is possible, however, that the placenta becomes more permeable to maternally produced conjugates in the later stages of gestation and this could account, in part, for DES conjugates appearing in the older fetuses.

Maternal brain tissue contained only unconjugated radioactivity, which indicates that the blood–brain barrier is also impermeable to polar conjugates of [^{14}C] DES. Brain contained a higher concentration of unconjugated radioactivity than plasma and much higher than the fetus. It appears that unconjugated DES and its metabolites cross the blood–brain barrier more rapidly than their passage across the maternal–fetal barrier. Goldstein et al. (1968) have calculated that such results may be due to a limiting blood flow to the placenta.

The maternal liver tissue contained tenfold higher concentrations of radioactivity than those found in the plasma. Only a minor fraction of hepatic tissue radioactivity was in the unconjugated form. This localization of conjugates in hepatic tissue reflects the importance of this organ as the site for metabolism and excretion of DES.

In addition to direct conjugation to glucuronic acid, DES undergoes oxidative metabolism in the liver, probably hydroxylation (Levin et al., 1968). These metabolites are slightly more polar than DES and are conjugated prior to their excretion in the bile. The unconjugated metabolites of DES, being relatively lipid soluble, could be expected to cross the placental barrier. To accurately determine how much unchanged DES appeared in the fetus, rats on day 13 of pregnancy were injected with 30 μCi [^{3}H] DES (Amersham-Searle, 6.2 Ci/mmol) in a manner identical to that described for [^{14}C] DES. The fetuses from each animal were combined and extracted with methanol as described. A reverse isotope dilution analysis of the fetal extracts for unchanged DES was performed. Addition of [^{3}H] DES to a blank (undosed) fetal extract showed that essentially all of the added radioactivity could be recovered in the form of DES (Table 2). Ten minutes after a dose of [^{3}H] DES to the mother 56.3% of the radioactivity in the fetus was unchanged DES. The percentage of fetal radioactivity that recrystalized from benzene with DES decreased with

TABLE 2. Unchanged DES in the Fetus after an Intravenous Dose of [³H]DES
(0.2 mg/kg) to Pregnant Rats on Day 13 of Gestation

Minutes after [³H]DES dose	Percent of fetal radioactivity as DES[a]	Percent of maternal dose in each fetus as DES
0	90.7[b]	—
10	56.3	0.0038
60	34.3	0.0015
90	27.6	0.0006

[a]Results of a reverse isotope dilution analysis for DES in 10 pooled fetuses from a single pregnant rat.

[b]This value was obtained by adding a known amount of [³H]DES to a blank fetal extract and conducting a reverse isotope dilution analysis.

time after the dose, and it could be calculated that only 0.0006% of the dose of DES remained in each fetus as unchanged drug at 90 min.

The results show that a placental barrier exists that excludes water-soluble conjugates and partially excludes unconjugated DES and its metabolites from reaching the fetus. In this respect DES is apparently different from unconjugated natural estrogens, which have been reported to cross to the fetus very rapidly (Bengtsson and Ullberg, 1963). Although the fetal concentrations of DES are lower than those in maternal tissues, it would seem reasonable that the concentrations in developing tissues containing estrogen-binding macromolecules could be higher than those observed in the whole fetus. It remains to be determined whether a similar pattern of maternal–fetal transfer of DES exists in the human and whether the predictably low concentrations that may be found could exert detrimental effects.

Absorption and Excretion of DES and DES Monoglucuronide in Humans

Because most animal species, including ruminants, rapidly metabolize DES by conjugation with glucuronic acid, animal tissues can contain DES primarily in the conjugated form (Huber et al., 1972). Diethylstilbestrol monoglucuronide (DESG) is converted to DES in the intestinal contents of laboratory animals, and the possibility exists that the estrogen could be released in the human intestine following ingestion of DESG. Thus, there is basis for concern because the glucuronide of DES, which is relatively inactive as an estrogen (Umberger, 1963), could be converted to a potentially carcinogenic substance in the human.

No information is available on the fate of DES or its glucuronide conjugate after oral ingestion by humans. This is a report of a limited study in humans conducted to determine whether it is possible for DES and/or its metabolites to appear in the blood after ingestion of DESG. In addition, indirect evidence was sought to implicate conjugate hydrolysis as an important step in the biological disposition of DESG. Two normal male subjects ingested $[^3H]$DES and $[^{14}C]$DES monoglucuronide ($[^{14}C]$DESG) simultaneously, and the plasma levels and excretion of each labeled material were followed.

The $[^{14}C]$DESG was prepared by isolation from the urine of a rabbit given 50 μCi $[^{14}C]$DES intraperitoneally (Fischer and Millburn, 1970). The conjugate was purified by Amberlite XAD-2 column chromatography (Bradlow, 1968) followed by preparative paper chromatography (Whatman 3MM), using the solvent system of Fewster and Hall (1951). Radioactive DES and DESG were chromatographically pure using previously described procedures (Fischer and Millburn, 1970). In addition, the purity of $[^{14}C]$DESG was confirmed by its complete hydrolysis using β-glucuronidase followed by reverse isotope dilution analysis for DES. After a constant specific activity was attained by recrystallization of carrier DES, the diacetate derivative was prepared and recrystallized to constant specific activity. These procedures showed that $[^{14}C]$DESG was 96% pure. No evidence of tritium exchange from $[^3H]$DES or its metabolites was found when biological samples were analyzed for ^3H_2O by dehydration followed by combustion analysis. Details of the protocol used for the administration of DES and DESG to humans are given in Fig. 6.

Simultaneous ingestion of $[^3H]$DES and $[^{14}C]$DESG allowed a direct comparison of their plasma levels and excretion kinetics. $[^3H]$DES appeared to be rapidly absorbed from the proximal portion of the gastrointestinal tract since peak concentrations of radioactive material in the plasma occurred between 20 and 40 min after ingestion (Fig. 6). In contrast, radioactive material from $[^{14}C]$DESG reached much lower peak plasma concentrations between 3 and 6 hr after ingestion. The difference in the time of peak plasma levels of ^{14}C and ^3H suggest that DESG was absorbed more slowly than DES.

DES was rapidly metabolized to polar metabolites in humans since no ether-extractable radioactive materials, such as DES or its unconjugated metabolites, could be detected in the plasma. However, up to 15% of the radioactive materials in the plasma could be DES since amounts below that figure were beyond the limits of detectability by our methods.

Excretion of radioactive material in the urine was essentially completed by 35 hr after ingestion of $[^3H]$DES and 60 hr after $[^{14}C]$DESG (Fig. 6). Slightly more of the radioactivity from $[^{14}C]$DESG was excreted in the urine (47%) than radioactivity from $[^3H]$DES (38%). The remainder of the radioactive dose was excreted in the feces over a 3-day period (Fig. 6). Unabsorbed DES and DESG, or the biliary excretion of these substances and their metabolites, could account for the radioactive materials in the feces.

The rates of excretion of radioactive materials in the urine following ingestion of $[^3H]$DES and $[^{14}C]$DESG were different as shown in Fig. 6. More tritiated materials than ^{14}C-labeled materials accumulated in the

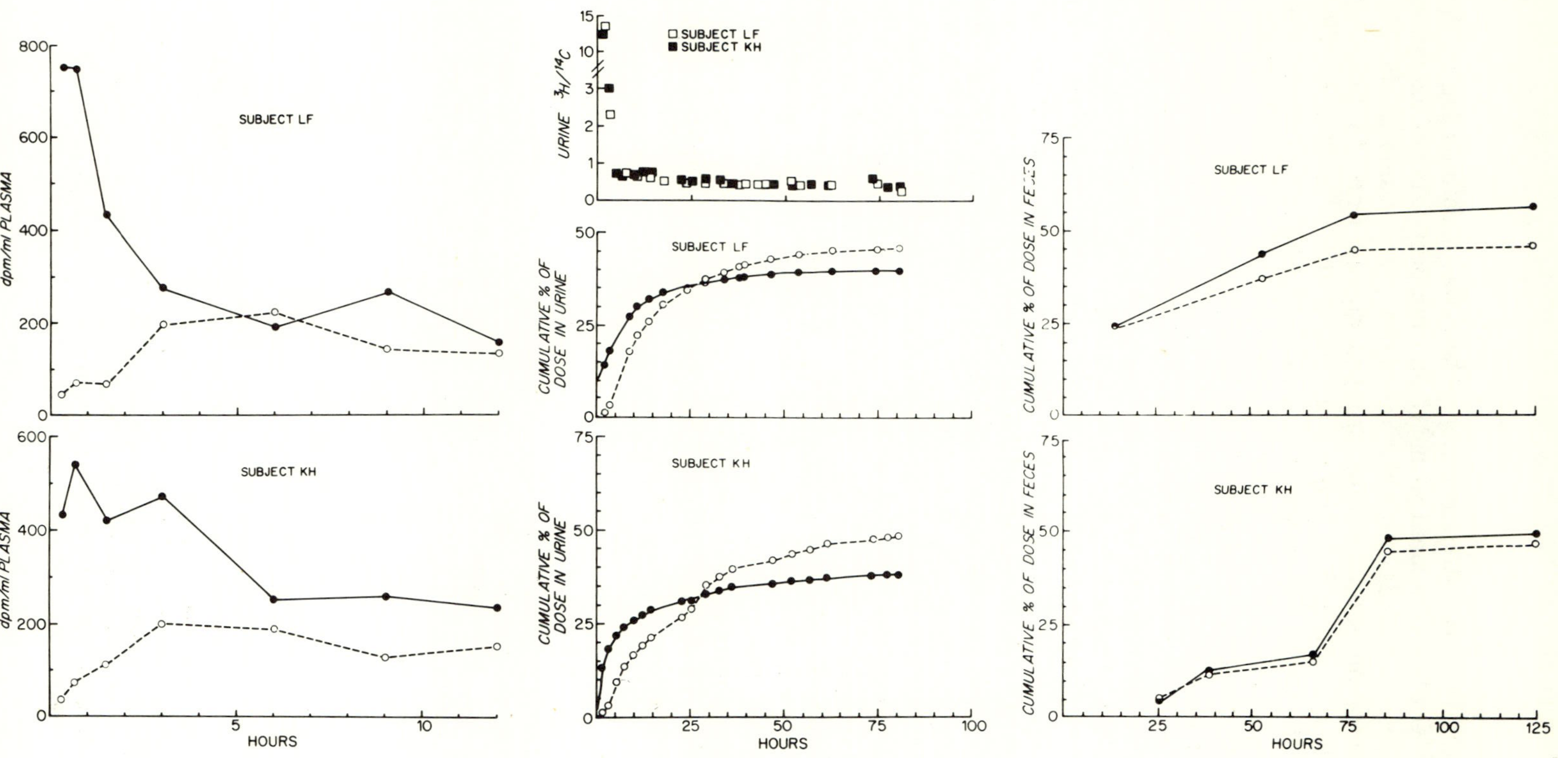

FIGURE 6. Plasma levels (left panel), urinary excretion (middle panel), and fecal excretion (right panel) of total radioactivity after ingestion of [³H]DES and [¹⁴C]DESG by two human volunteers. Closed circles (●) represent tritiated materials; open circles (○) represent ¹⁴C-labeled materials. Open and closed squares (top, middle panel) represent the ratio of the concentrations of ³H and ¹⁴C in the urine at various times.

Two male volunteers (LF, age 36, wt. 70 kg and KH, age 24, wt. 70 kg) ingested radiolabeled DES and DESG after an overnight fast. [³H]DES (5.0 μCi, 0.26 μg) and DESG(monoethyl-1-¹⁴C) (5.0 μCi, 60μg) were dissolved in 5 ml of ethanol and administered together in a hydroalcoholic solution (150 ml). Blood samples were obtained over a 24-hr period, and urine and feces were quantitatively collected for 6 days. Aliquots of plasma, urine, and fecal homogenates were analyzed for ³H and ¹⁴C separately by combustion in a Packard tritium carbon oxidizer followed by liquid scintillation.

urine during the first day after ingestion. There was a definite lag in the excretion of ^{14}C when compared with ^{3}H, and the peak rates of excretion of both isotopes coincided with their respective peak plasma levels. The ^{3}H:^{14}C concentration ratio in urine dropped rapidly from 13 (ingested dose = 1.18) to 0.5 within 5 hr and did not change thereafter (Fig. 6). This indicates that the major differences that exist in the biological disposition of DES and its glucuronide conjugate occurred during the first 5 hr after ingestion. This transient difference is probably in absorption from the gastrointestinal tract since a difference in excretion or metabolism to more rapidly excreted metabolites would cause a continual change in the ^{3}H:^{14}C ratio until metabolism and/or elimination were complete.

The radioactive products in the urine were characterized to determine whether there were differences in the metabolic products of [^{3}H]DES and [^{14}C]DESG. The results in Table 3 indicate that no major differences existed between the ^{3}H-labeled and ^{14}C-labeled radioactive materials in the urine when pooled samples were subjected to enzyme hydrolysis and solvent extraction. Humans excreted very little nonpolar material in the urine, and the majority of the metabolic products could be characterized as glucuronide conjugates. Less than 16% of the radioactive products in urine were sulfates. A slightly larger fraction was characterized as polar materials that could not be hydrolyzed by glucuronidase or sulfatase. These data are consistent with the urinary metabolite pattern of DES in laboratory animals (Hanahan et al., 1953; Hinds et al., 1965).

Paper chromatography was used to further characterize the urinary metabolites of [^{3}H]DES and [^{14}C]DESG. Pooled urine samples (60–100

TABLE 3. Characterization of Radioactive Materials in Urine after Oral Ingestion of [^{14}C]DESG and [^{3}H]DES by Two Human Subjects

| | | Percent of urinary radioactivity characterized[a] as | | | | | | | |
| | | Unconjugated | | Glucuronides | | Sulfates | | Polar | |
Subject	Urine pool (hr)	^{3}H	^{14}C	^{3}H	^{14}C	^{3}H	^{14}C	^{3}H	^{14}C
LF	0–24	0	4	67	80	10	6	24	10
	24–48	8	10	81	63	12	14	0	13
KH	0–24	2	2	60	80	9	15	29	4
	24–48	3	4	58	67	7	16	32	12

[a]Aliquots (0.5 ml) of pooled urine were adjusted to pH 7.0 and extracted with 10 volumes of ether. The unconjugated fraction represents ether-extractable materials in urine. The glucuronide metabolite fraction represents ether-extractable materials after a 16-hr incubation at 37°C with β-glucuronidase (1 ml Ketodase-Warner Chilcott). The sulfate fraction was determined by ether extraction after hydrolysis of urine with a β-glucuronidase-aryl sulfatase mixture (Glusulase-Endo). The increase in extractable radioactive materials over those released by β-glucuronidase alone represents sulfate conjugates. The polar fraction represents radioactive metabolites that are not extracted into ether after enzyme hydrolysis.

ml) containing 50 μg each of added carrier DES and DESG were partially purified using a column containing Amberlite XAD-2 (Bradlow, 1968). Radioactive metabolites were completely eluted from the column and applied to Whatman No. 1 paper for chromatography. The chromatographic solvent systems employed have been previously described (Fischer and Millburn, 1970). An n-propanol:NH_4OH (7:3) system gave the best separation of metabolites as determined by combustion analysis of consecutive 2-cm portions of the developed chromatograms. Essentially, the same chromatographic pattern was seen in 0–24-hr and 24–48-hr urine from each subject. Approximately 50% of the ^{14}C-labeled material in the urine migrated in a manner identical to authentic DESG ($R_f = 0.7$). The remainder of the urinary ^{14}C-labeled metabolites migrated as a single spot with an R_f of 0.4. An identical chromatographic pattern was seen for ^{3}H-labeled metabolites of DES except that a smaller percentage of the radioactive materials (~30%) were chromatographically identical to DESG. No evidence of [3]H- or ^{14}C-labeled DES ($R_f = 0.90$) in the urine was found using these procedures. The fact that about half the ^{14}C-labeled material in the urine after ingestion of [^{14}C]DESG was conjugated DES was confirmed by reverse isotope dilution analysis. An aliquot of pooled urine (0–24 hr) from each subject was subjected to β-glucuronidase hydrolysis followed by extraction of the released aglycones into ether. A known amount of carrier DES was added to the extract and subsequently recrystallized from benzene to constant specific activity. It was found that 46–49% of the extracted radioactive materials represented DES in the two samples analyzed. Since this DES was present in the urine conjugated to glucuronic acid, it can be calculated that 48% of the total ^{14}C-labeled material in the urine was DESG.

The results of the enzyme hydrolysis and chromatographic procedures used to characterize the metabolites excreted in the urine after [^{3}H]DES and [^{14}C]DESG ingestion suggest that each substance was converted to the same metabolic products in the body. Since the ingested glucuronide conjugate was excreted as products other than DESG, it appears that conjugate hydrolysis occurs in the body. Hydrolysis of DESG to DES may be nearly complete, since similar amounts of sulfate conjugates and polar nonhydrolyzable metabolites were excreted in the urine after ingestion of DES and its glucuronide conjugate (Table 3).

Information concerning the disposition of DES and its glucuronide conjugate in experimental animals and the fate of conjugates of natural estrogens in humans facilitate interpretation of our results. Animal studies show that DES is rapidly absorbed, which appears to be the case for humans. Assuming nearly complete absorption, the fecal elimination of the drug can be attributed to biliary excretion. Most estrogens, including DES, are excreted in large amounts in the bile of humans and laboratory animals as polar metabolites, usually glucuronides (Hinds et al., 1965; Jayle and Pasqualini, 1966; Sandberg et al., 1970).

The monoglucuronide of DES and other glucuronide conjugates (Cottle and Veress, 1971; Herz et al., 1961; Lester and Schmid, 1963; Schacter et al., 1959) are not well absorbed in their intact form from the gastrointestinal tracts of laboratory animals. Appearance of these conjugates in the intestine, usually via their excretion in bile, is followed by hydrolysis of the glucuronides by β-glucuronidase, an enzyme produced by bacteria that inhabit the lower intestine (Kent et al., 1972). The aglycones resulting from hydrolysis are usually more readily absorbed than the conjugates. For example, it has been shown in this report that introduction of DESG into the lower intestine of anesthetized rats resulted in rapid hydrolysis of the conjugate to DES followed by absorption of the unconjugated drug. The excretion data show that DESG probably meets a similar fate in the intestines of humans. The lag time in attainment of peak plasma levels and urinary excretion rates of ^{14}C-labeled materials after ingestion of [^{14}C]DESG suggest that movement of the conjugate down the gastrointestinal tract to an area high in β-glucuronidase activity may precede conjugate hydrolysis and absorption of the liberated DES. The attainment of a constant ^{3}H:^{14}C ratio in the urine after ingestion of [^{3}H]DES and its ^{14}C-labeled conjugate is consistent with hydrolysis of the conjugate to [^{14}C]DES. Also consistent with intestinal hydrolysis of DESG is the fact that over 80% of the labeled materials excreted in the feces after [^{14}C]DESG ingestion were soluble in ether. Conjugated materials in the feces would not have partitioned into ether from a neutral aqueous solution. One could speculate, however, that DESG was absorbed intact from the intestine and hydrolyzed elsewhere in the body. This appears unlikely since glucuronide conjugates of other estrogens, given intravenously to humans, are excreted in the urine primarily in the intact form if their entrance into the intestine via the bile is prevented (Sandberg et al., 1970).

The elimination in the feces of a large portion of the ingested DES and DESG makes estimation of the exact amount of each material absorbed difficult. At a minimum, half of the ingested DESG was absorbed in the two subjects since 50% of the ingested radioactivity was excreted in the urine. Indirect evidence showed that the conjugate must have been hydrolyzed in the intestinal tract. Since large differences between sexes have not been found for the disposition of foreign compounds in humans (O'Malley et al., 1971; Vesell and Page, 1968a, b), it appears that ingestion of DESG by males or females can provide plasma levels of circulating DES and/or its metabolites. Whether there are adverse effects attributable to the low levels of these materials in the body that may result from their ingestion as residues in meat remains unknown.

CONCLUSIONS

The disposition of DES in laboratory animals and humans does not appear to be dramatically different from that of the natural estrogens.

Metabolism of the drug takes place in the intestine and liver with glucuronide conjugates of the drug and its metabolites undergoing excretion in the bile. The polar conjugates, after biliary excretion, are not well absorbed from the intestine but are hydrolyzed by β-glucuronidase in the distal portion of the intestine. Absorption of the released aglycones occurs, but binding to intestinal contents slows the absorption process. It is not known whether this enterohepatic cycling can actually prolong peripheral levels of DES and its metabolites. The reabsorbed drug and metabolites are cleared rapidly by the liver so that very little may escape metabolism and biliary excretion to reach peripheral target tissues. Hepatic clearance of DES may also be rapid in man since unconjugated radioactive materials in the plasma of humans were not detected after a dose of $[^3H]$DES in the studies reported here.

Experiments in newborn rats indicated that two important processes involved in the enterohepatic circulation of DES developed after birth. Both conjugation to glucuronic acid and deconjugation in the intestine reached adult levels by the 25th day after birth. The age at which the human infant reaches an adult capacity to handle DES is not known. From evidence available for other drugs one could postulate that the human infant would be deficient in these processes at birth. Infants are known to conjugate endogenous and exogenous compounds with glucuronic acid more slowly than the adult (Dutton, 1966), and Kent et al. (1972) showed that the human may not have adult levels of intestinal β-glucuronidase activity at the age of 1 yr. Diminished hydrolysis of biliary conjugates in the intestine, thereby reducing the enterohepatic cycling of aglycones, may act as a protective mechanism to reduce exposure of young animals to possibly harmful substances.

Other studies have also shown that administration of DES by parenteral routes will produce a greater uterotropic effect than that produced by the same oral dose (Umberger, 1963). Results presented in this report show that the ability of phenobarbital, a known inducer of drug metabolism, to decrease the levels of DES in the uterus and reduce its effect on that organ also depends on the route of estrogen administration. Phenobarbital pretreatment had effects when the estrogen was given orally but not when given parenterally. Since DES is given orally for therapeutic purposes or possibly ingested as a residue in meat, this observation may be of obvious practical importance.

It was shown in these studies that DES crossed the rat placenta after maternal administration. Total fetal tissue contained low concentrations of unconjugated DES and metabolites; it can be concluded that the drug was restricted in its movement across the placenta. Water-soluble DES metabolites did not reach the fetus at midgestation, which is consistent with the placenta acting as a lipoid barrier. These results should be of value in the design and evaluation of experiments on the possible harmful effects caused by the administration of DES to pregnant animals.

Comparison of plasma levels and excretion of radiolabeled products after the ingestion of [^{3}H]DES and [^{14}C]DESG by two human subjects suggested that the fate of each substance followed a pattern similar to that observed in laboratory animals. The absorption of DES was apparently more rapid than the absorption of DESG. Indirect evidence indicated that after ingestion of DESG, the conjugate was hydrolyzed in the intestinal tract. From urinary excretion data, it was estimated that at least half the ingested DESG was absorbed, probably as unconjugated DES, from the gastrointestinal tract of human subjects. These results are the first available on the quantitative aspects of the biological disposition of DES and its monoglucuronide conjugate in humans. To help evaluate the potential hazardous effects after ingestion of DES and its metabolites as residues in meat, these experiments should obviously be expanded to include the ingestion of the products as they are found in prepared food.

QUESTIONS AND ANSWERS

W. D. Price: Would you comment on the differences in the distribution and metabolism of DES in various species including man?

L. J. Fischer: There are no dramatic differences in the distribution of estrogens between man and a number of laboratory animals. DES distribution and metabolism are very similar in humans, mice, and rats. The drug is excreted in the bile and undergoes enterohepatic circulation.

S. R. Cernosek: What constituted the human population used in these studies?

L. J. Fischer: Two normal male subjects were used, and we are currently conducting further studies in another male. We have no reason to believe that DES distribution and metabolism in males should be different from females. In rats sex differences in metabolism have been observed for many xenobiotics, but this difference does not appear to exist in humans.

REFERENCES

Bengtsson, G. and Ullberg, S. 1963. The autoradiographic distribution pattern after administration of diethylstilboestrol compared with that of natural estrogens. *Acta Endocrinol.* 43:561–580.

Bradlow, H. L. 1968. Extraction of steroid conjugates with a neutral resin. *Steroids* 11:265–272.

Cottle, W. H. and Veress, A. T. 1971. Absorption of the glucuronide conjugate of triiodothyronine. *Endocrinology* 88:522–523.

Diczfalusy, E., Tillinger, K. G., Wiqvist, N., Levitz, M., Condon, G. P. and Dancis, J. 1963. Disposition of intra-amniotically administered estriol-16-C^{14} and estrone-16-C^{14} sulfate by women. *J. Clin. Endocrinol. Metab.* 23:503–509.

Dutton, G. J. 1966. The biosynthesis of glucuronides. In *Glucuronic acid free and combined*, ed. G. J. Dutton, p. 263. New York: Academic Press.

Fewster, M. E. and Hall, D. A. 1951. Application of buffered solvent systems to the detection of aromatic acids by paper partition chromatography. *Nature* 168:78–79.

Fischer, L. J. and Millburn, P. 1970. Stilboestrol transport and glucuronide formation in everted sacs of rat intestine. *J. Pharmacol. Exp. Ther.* 175:267–275.

Fischer, L. J. and Weissinger, J. L. 1972. Development in the newborn rat of the conjugation and deconjugation process involved in the enterohepatic circulation of diethylstilbestrol. *Xenobiotica* 2:399–412.

Fischer, L. J., Millburn, P., Smith, R. L. and Williams, R. T. 1966. The fate of ^{14}C-stilboestrol in the rat. *Biochem. J.* 100:69P.

Fischer, L. J., Kent, T. H. and Weissinger, J. L. 1973. Absorption of diethylstilbestrol and its glucuronide conjugate from the intestines of five- and twenty-five day-old rats. *J. Pharmacol. Exp. Ther.* 185:163–170.

Goebelsmann, U., Wiqvist, N., Diczfalusy, E., Levitz, M., Condon, G. P. and Dancis, J. 1966. Fate of intra-amniotically administered oestriol-15-^{3}H-3-sulphate and oestriol-16-^{14}C-16-glucosiduronate in pregnant women at midterm. *Acta Endocrinol.* 52:550–564.

Goldstein, A., Aranow, L. and Kalman, S. M. 1968. *Principles and drug action*, p. 160. New York: Harper & Row.

Greenwald, P., Barlow, J. J., Nasca, P. C. and Burnett, W. S. 1971. Vaginal cancer after maternal treatment with synthetic estrogens. *New Engl. J. Med.* 285:390–392.

Hanahan, D. J., Daskalakis, E. G., Edwards, T. and Dauben, H. J. 1953. The metabolic pattern of ^{14}C-diethylstilbestrol. *Endocrinology* 53:163–170.

Herbst, A. L., Ulfelder, H. and Poskanzer, D. C. 1971. Adenocarcinoma of the vagina, association of maternal stilbestrol therapy with tumor appearance in young women. *New Engl. J. Med.* 284:878–881.

Herz, R., Tapley, D. F. and Ross, J. E. 1961. Glucuronide formation in the transport of thyroxine analogues by rat intestine. *Biochim. Biophys. Acta* 53:273–284.

Hinds, F. C., Draper, H. H., Mitchell, G. E. and Neumann, A. L. 1965. Metabolism of labeled diethylstilbestrol in ruminants. *J. Agric. Food Chem.* 13:256–259.

Huber, T. L., Horn, G. W. and Beadle, R. E. 1972. Liver and gastrointestinal metabolism of diethylstilbestrol in sheep. *J. Animal Sci.* 34:786–790.

Jayle, M. F. and Pasqualini, J. R. 1966. Implication of conjugation of endogenous compounds—steroids and thyroxine. In *Glucuronic acid free and combined*, ed. G. J. Dutton, p. 530. New York: Academic Press.

Kent, T. H., Fischer, L. J. and Marr, R. 1972. Glucuronidase activity in intestinal contents of rat and man and relationship to bacterial flora. *Proc. Soc. Exp. Biol. Med.* 140:590–594.

Klaassen, C. D. 1973. The effect of altered hepatic function on the toxicity, plasma disappearance and biliary excretion of diethylstilbestrol. *Toxicol. Appl. Pharmacol.* 24:142–149.

Lester, R. and Schmid, R. 1963. Intestinal absorption of bile pigments 1. The enterohepatic circulation of bilirubin in the rat. *J. Clin. Invest.* 42:736–746.

Levin, W., Welch, R. M. and Conney, A. H. 1968. Decreased uterotrophic potency of oral contraceptives in rats pretreated with phenobarbital. *Endocrinology* 83:149–156.

Levitz, M. 1966. Conjugation and transfer of fetal-placental steroid hormones. *J. Clin. Endocrinol. Metab.* 26:773–777.

O'Malley, K., Crooks, J., Duke, E. and Stevenson, I. H. 1971. Effect of age and sex on human drug metabolism. *Brit. Med. J.* 3:607–609.

Sandberg, A. A., Slaunwhite, W. R. and Kirdani, R. Y. 1970. Enterohepatic circulation of estrogens. In *Metabolic conjugation and metabolic hydrolysis*, ed. W. H. Fishman, vol. II, pp. 138, 140. New York: Academic Press.

Schacter, D., Kass, D. J. and Lannon, T. J. 1959. The biosynthesis of salicyl glucuronides by tissue slices of various organs. *J. Biol. Chem.* 234:201–205.

Umberger, E. J. 1963. Relative uterotrophic potency of diethylstilbestrol and diethylstilbestrol monoglucuronide in rats and mice. *Endocrinology* 73:650–652.

Vesell, E. S. and Page, J. G. 1968a. Genetic control of drug levels in man: Phenylbutazone. *Science* 159:1479–1480.

Vesell, E. S. and Page, J. G. 1968b. Genetic control of drug levels in man: Antipyrine. *Science* 161:72–73.

Received June 4, 1975
Accepted June 27, 1975

ISOLATION AND IDENTIFICATION OF METABOLITES OF ^{14}C-LABELED ESTRADIOL IN CATTLE

C. C. Kaltenbach, T. G. Dunn, D. R. Koritnik, W. F. Tucker

Division of Animal Science, University of Wyoming, Laramie, Wyoming

D. B. Batson

Food and Drug Administration, Rockville, Maryland

R. B. Staigmiller, G. D. Niswender

Department of Physiology and Biophysics, Colorado State University,
Fort Collins, Colorado

A total of six steers and six heifers received three daily injections containing either 200 μCi (1 mg) of [4-^{14}C]estradiol-17β or 312 μCi (2.16 mg) of [4-^{14}C]estradiol-17β 3-benzoate. Major metabolites of the administered estradiol-17β and estradiol-17β 3-benzoate were identified in muscle, fat, liver, and kidney samples obtained 3 hr after the final injection.

Estradiol benzoate was nondetectable in the tissues analyzed, suggesting rapid hydrolysis of the benzoate ester. Consequently, the relative proportions of the various metabolites were similar for both the injected estrogens.

Estradiol-17β and estrone, which together accounted for 80-90% of the total extracted radioactivity, appear to be the major metabolites in both muscle and fats. In contrast, the major metabolites present in liver and kidney appear in the conjugate fraction. Most of the conjugated metabolites were glucuronates, which represent 85–95% of the total recovered conjugate radioactivity.

INTRODUCTION

Widespread use of various estrogens as growth stimulants in meat animals could lead to an accumulation of these compounds or their metabolites in the human diet. Little is known about the metabolic fate of these compounds. The objectives of the study reported here were to isolate and identify the major metabolites of administered estradiol-17β and estradiol-17β 3-benzoate in tissues of steers and heifers.

METHODS

A total of six steers and six heifers, weighing approximately 200 kg, were used in this study. Three steers and three heifers each were assigned to receive either estradiol-17β or estradiol benzoate.

Requests for reprints should be sent to C. C. Kaltenbach, Division of Animal Science, University of Wyoming, Laramie, Wyoming 82071.

Journal of Toxicology and Environmental Health, 1:607–616, 1976

All animals received a subcutaneous injection of 1 mg of the appropriate estrogen in propylene glycol–peanut oil (30:70) at the base of the ear once daily for 11 days. The injections of nonradioactive estrogen were followed by three daily injections of [^{14}C]estrogen in the same carrier (either 1 mg [4-^{14}C]estradiol-17β or 2.16 mg [4-^{14}C]estradiol-17β 3-benzoate). Specific activity of each estrogen was 54.5 mCi/mmol resulting in radioactive doses of 200 μCi of estradiol-17β or 312 μCi of estradiol benzoate. A preliminary study indicated it was necessary to give a larger dose of the radioactive estradiol benzoate than of estradiol-17β to achieve comparable blood levels for both hormones and that peak levels of radioactivity occurred in the blood approximately 3 hr after injection.

Accordingly, animals were sacrificed 3 hr after the third injection of radioactive estrogen. The following samples were collected and frozen on Dry Ice: neck muscle, rump muscle, neck fat, rump fat, mesenteric fat, liver, kidney, and kidney fat. The entire liver (following removal of the gall bladder) was immediately homogenized in a Waring blender, and 400 g of the homogenate was frozen. All samples were stored at $-16°$C until analysis.

Subsamples of all tissues and fluids were solubilized to allow determination of total radioactivity. Three subsamples of each tissue, weighing approximately 50, 100, and 150 mg, were prepared by weighing thin slices of the frozen tissue directly into glass liquid scintillation counting (LSC) vials.

All tissues except fat were digested by adding 1.0 ml of 2 N NaOH to the weighed tissue in the glass vial and heating the tightly capped vial at 130°C for 30 min in a sand bath. After cooling, 2 drops of freshly prepared 14% ascorbic acid and 2.5 ml of neutralizing solubilizer (Beckman BBS-2) were added. The contents of the vial were vigorously mixed on a vortex mixer. Ten minutes later, 10 ml toluene LSC fluid (3 g PPO + 0.1 g POPOP in 1 liter of toluene) were added and the contents of the vial again thoroughly mixed.

Fat was digested in vials with 1.0 ml of 0.2 N NaOH at 130°C for 30 min. After cooling, 5 drops of 10% acetic acid and 2 drops of freshly prepared 14% ascorbic acid were added. After vigorous mixing, 10 ml of LSC–solubilizer (toluene cocktail–Beckman BBS-3, 4:1) were added and the contents of the vial thoroughly mixed. The fat samples were allowed to stand in the dark for 24 hr before counting.

The extraction scheme for tissues is shown in Fig. 1. Frozen tissue was minced into a 50-ml stainless steel Omnimixer homogenizing chamber containing 100 μg each of estradiol-17β, estradiol-17α, estradiol-17β 3-benzoate, estrone, estriol, estrone glucuronate, estradiol glucuronate, and estradiol sulfate in 25 ml absolute ethanol. These steroids were added to aid in the isolation and identification procedures. The tissue was homogenized for 1 min at 15,000 rpm. The residue was stripped from the impeller and shaft, and a second 1-min homogenization was carried out without changing solvent. The chamber was kept immersed in an ice-water bath during homogenization. The homogenizing chamber and its contents were centri-

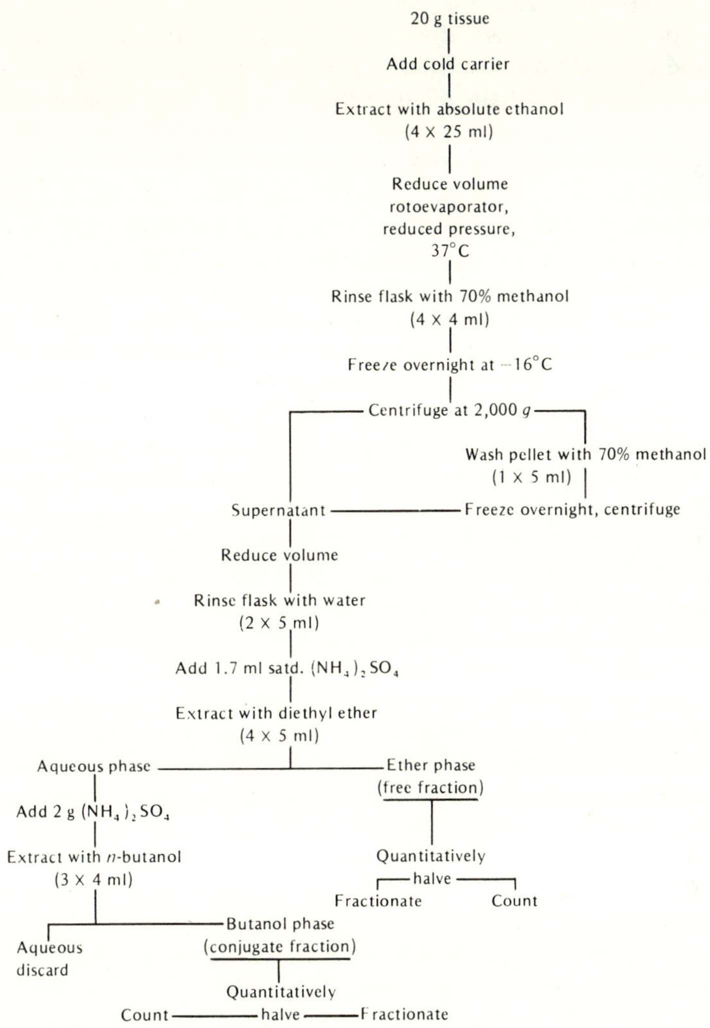

FIGURE 1. Flowchart for extraction of radioactive metabolites from tissues and fluids.

fuged at 2000 g for 10 min (4°C) and the supernatant decanted. The pellet was homogenized and centrifuged three additional times, and the supernatant fractions were pooled and reduced to dryness. The extract was then transferred to a 40-ml conical centrifuge tube with four, 4-ml washes of 70% methanol. The samples were stored overnight at −16°C then centrifuged at 2000 g for 10 min at 4°C and the supernatant decanted. The pellet was resuspended in 5 ml of 70% methanol and centrifuged again. This procedure removed most of the nonsteroidal lipids. The supernatants were pooled and reduced to dryness; the residue was transferred with two 5-ml water washes to a 60-ml separatory funnel. Saturated $(NH_4)_2SO_4$ (1.7 ml) was added and this mixture was extracted four times with 5 ml diethyl ether. The ether extracts, which contained the free estrogens, were pooled and reduced to dryness. The residues were dissolved in absolute methanol, quantitatively divided, and one-half was counted to establish procedural losses.

The remaining one-half of the free extract was applied to an LH-20 (Sephadex) column (1 X 15 cm) and eluted with benzene:methanol (85:15), according to the procedure of Carr et al.(1971). The first 14 ml of eluate contained estrone (E_1) and estradiol benzoate (E_2 benzoate); estradiol (E_2) eluted between 15 and 27 ml; and estriol (E_3) eluted between 28 and 50 ml.

The E_3 fraction was evaporated and counted. The E_1 and E_2 benzoate fraction was evaporated and spotted on prepared silica gel TLC plates (Brinkman). The plates were developed with benzene:methanol (85:15), air dried, and visualized with Barton's reagent (3% ferric chloride:3% potassium ferricyanide:water, 1:1:3). The E_2 benzoate and E_1 regions were cut out and eluted into counting vials with 10 ml of benzene:methanol (85:15). The eluate was dried and counted. The E_2 fraction was evaporated and spotted on Whatman No. 1 paper strips that had been saturated in formamide: methanol (1:1) and blotted. These strips were developed by descending chromatography with 20 ml per strip of hexane:benzene (1:1) saturated with formamide (Zaffaroni, 1953). The strips were air dried and sprayed with Barton's reagent. The estradiol-17β ($E_2\beta$) and estradiol-17α ($E_2\alpha$) bands were cut out and eluted into counting vials with 20 ml of methanol. The eluate was dried and counted.

After extraction of the free estrogens with ether, 2 g of $(NH_4)_2SO_4$ were added to the remaining aqueous phase to enhance extraction of the conjugated estrogens. This mixture was extracted three times with 4 ml of *n*-butanol. The *n*-butanol extracts, which contained the conjugated estrogens, were pooled and taken to dryness.

One-half the conjugate was counted in LSC–solubilizer to establish procedural losses. Samples of the conjugate extract containing more than 400 dpm/10 g tissue were applied to an LH-20 column (1 X 22 cm) and eluted with chloroform:methanol (1:1) in 0.01 M NaCl (Vihko, 1966). The first 27 ml contained the glucuronates and the 28- to 54-ml fraction contained the sulfates. One-half of each fraction was counted to determine recoveries (Fig. 2).

The second half of the glucuronate fraction was hydrolyzed with β-glucuronidase (bacterial, type I, Sigma Chemical). The glucuronate fraction was mixed with 5 ml of phosphate buffer (pH 7.0) and 1 drop of chloroform, and approximately 500 Fishman units of enzyme were added to the buffered mixture. The entire mixture was incubated for 5 days at 37°C. After enzyme hydrolysis the mixture was partitioned and worked up using the standard procedure described above for "defatted" tissue extracts. To monitor the efficiency of hydrolysis, the aqueous phase was again extracted with *n*-butanol and the extract counted. Enzyme hydrolysis releases 60–90% of the conjugate radioactivity to the free fraction. The remaining non-hydrolyzed fraction has not been characterized.

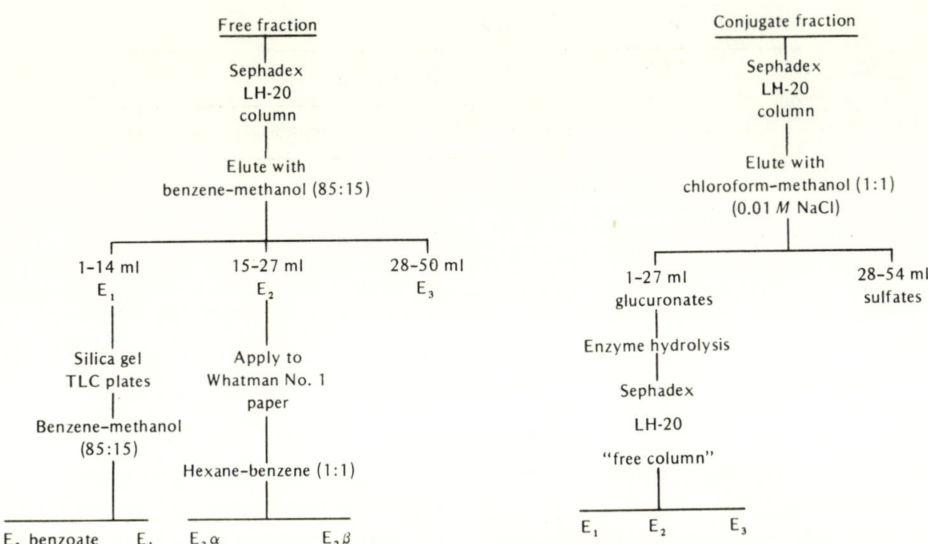

FIGURE 2. Fractionation scheme for estrogen metabolites extracted as in Figure 1.

The data were analyzed by least-squares analysis of variance for unequal subclass numbers (Harvey, 1960).

RESULTS AND DISCUSSION

Values obtained for total radioactivity in the various tissues collected from each animal are reported in Table 1. One animal in the estradiol-17β-treated group exhibited considerably lower levels of tissue radioactivity than the other animals in this group. Values for this animal, which deviated from the mean of the group in most instances by more than two standard deviations, were excluded from the data. It is theorized that the material injected into this animal may have been entrapped in cartilage, thus rendering it unavailable to the circulatory system.

Muscle, with no known estrogen requirement for normal activity, contained relatively low levels of radioactive estrogens. The observed radioactivity may have been due to the presence of blood or intramuscular fat rather than the muscle tissue proper. Fat tissue contains about three times the level of radioactivity found in muscle tissue. The higher levels of radioactivity in fat may be due to high solubility of the estrogens in lipid material. Radioactivity was significantly higher in liver and kidney than in muscle or fat. This is in agreement with the known metabolic functions of these organs.

The concentrations of tissue metabolites, calculated from tissue radioactivity and expressed as ng/g, are given in Table 2. Molecular weights of the isolated metabolites were utilized to calculate the total tissue concentrations. A molecular weight of 272 was assumed for calculation of free estrogens and 349 was used to calculate the conjugate fractions.

TABLE 1. Total Tissue Radioactivity[a]

Tissue		Estradiol-17β		Estradiol-17β 3-benzoate
	n	Mean ± SE	*n*	Mean ± SE
Rump muscle	5	1,646 ± 290	6	1,033 ± 142
Neck muscle	5	1,767 ± 303	6	1,488 ± 264
Rump fat	5	5,256 ± 702	6	3,447 ± 357
Neck fat	5	5,359 ± 717	5	3,493 ± 435
Mesenteric fat	5	8,695 ± 766	6	3,901 ± 433
Kidney fat	5	8,841 ± 1,305	6	4,339 ± 446
Kidney	5	17,719 ± 2,360	6	20,767 ± 5,895
Liver	5	23,898 ± 2,113	6	24,208 ± 6,137

[a]dpm/10 g.

Analysis of covariance indicated that body weight had no effect on tissue concentration of the various metabolites. Subsequent analysis of variance indicated there were no differences in the concentrations of total metabolites found in muscle or fat from the neck versus rump regions. This was true for both the estradiol and estradiol benzoate treated animals.

In mesenteric fat and kidney fat ($p < 0.01$) the concentrations of total metabolites were significantly lower in estradiol benzoate-treated animals than in estradiol-treated animals. A similar trend, although not significant ($p > 0.10$), was observed for the two muscle locations, rump fat, and neck fat. This trend was reversed, however, in kidney and liver.

Steers had significantly ($p < 0.05$) higher concentrations of total metabolites than heifers in neck muscle. In the remaining muscle and fat samples, steers tended to have higher concentrations of total metabolites than heifers, although the differences were not significant ($p > 0.10$). Again, this trend was reversed for kidney and liver.

A treatment X sex interaction occurred for rump fat and kidney fat ($p < 0.05$). These interactions occurred because steers treated with estradiol had higher concentrations of total metabolites than did steers treated with estradiol benzoate or heifers treated with either estradiol or estradiol benzoate. A similar but nonsignificant ($p > 0.10$) trend also appeared for neck muscle, neck fat, and mesenteric fat. In the liver, an opposite trend ($p > 0.10$) occurred where estradiol benzoate-treated heifers had higher levels than estradiol-treated heifers or steers treated with either estrogen.

The relative proportions of the various metabolites of each of the injected estrogens are summarized in Table 3. Estradiol-17β and estrone appear to be the major radioactive metabolites in both muscle and fat.

TABLE 2. Least-Squares Means and Standard Errors for Exogenous Radioactive Estrogens in Muscle, Fat, Kidney, and Liver[a]

Treatment and tissue	Steers	Heifers	Mean ± SE
Rump muscle[b]			
$E_2\beta$	0.45	0.26	0.36 ± 0.05^c
benzoate	0.22	0.25	0.23 ± 0.04^c
mean ± SE	0.34 ± 0.04	0.25 ± 0.05	0.29 ± 0.03
Neck muscle			
$E_2\beta$	0.51	0.24	0.37 ± 0.05
benzoate	0.39	0.28	0.34 ± 0.05
mean ± SE	0.45 ± 0.05^d	0.26 ± 0.05^d	0.36 ± 0.04
Rump fat			
$E_2\beta$	1.41^e	0.86^e	1.14 ± 0.10^c
benzoate	0.70^e	0.86^e	0.78 ± 0.09^c
mean ± SE	1.06 ± 0.09	0.86 ± 0.10	0.96 ± 0.06
Neck fat			
$E_2\beta$	1.40	0.94	1.17 ± 0.13^c
benzoate	0.76	0.81	0.79 ± 0.13^c
mean ± SE	1.08 ± 0.13	0.88 ± 0.13	0.98 ± 0.09
Mesenteric fat			
$E_2\beta$	2.14	1.72	1.93 ± 0.14^f
benzoate	0.79	0.98	0.89 ± 0.12^f
mean ± SE	1.47 ± 0.12	1.35 ± 0.14	1.41 ± 0.09
Kidney fat			
$E_2\beta$	2.37^e	1.46^e	1.92 ± 0.17^f
benzoate	0.84^e	1.13^e	0.99 ± 0.15^f
mean ± SE	1.61 ± 0.15	1.30 ± 0.17	1.45 ± 0.11
Kidney			
$E_2\beta$	5.34	4.13	4.74 ± 1.39
benzoate	3.72	7.84	5.78 ± 1.24
mean ± SE	4.53 ± 1.24	5.99 ± 1.39	5.26 ± 0.93
Liver			
$E_2\beta$	5.95	7.37	6.66 ± 1.28
benzoate	4.12	9.47	6.79 ± 1.15
mean ± SE	5.04 ± 1.15^g	8.42 ± 1.28^g	6.73 ± 0.86

[a]ng/g tissue.

[b]Numbers per subgroup were estradiol steers = 3; estradiol heifers = 2; estradiol benzoate steers = 3; and estradiol benzoate heifers = 3 except for neck fat where estradiol benzoate steers = 2.

[c]Estradiol treated greater than estradiol benzoate treated ($p < 0.10$).

[d]Steers greater than heifers ($p < 0.05$).

[e]Treatment × sex interaction ($p < 0.05$).

[f]Estradiol treated greater than estradiol benzoate treated ($p < 0.01$).

[g]Heifers greater than steers ($p < 0.05$).

TABLE 3. Means of Extracted Radioactivity for Individual Metabolites Expressed as a Percent of Total Recovered Radioactivity

Tissue	Free fraction					Conjugate fraction		
	Total	E_1	$E_2\alpha$	$E_2\beta$	E_3	Total	Glucuronates	Sulfate
Estradiol-17β								
rump muscle	95	22	2	67	4	4		
neck muscle	95	21	3	64	5	5		
rump fat	98	27	2	66	2	2		
neck fat	96	27	3	54	9	4		
mesenteric fat	95	32	3	50	6	6		
kidney fat	95	25	2	57	7	6		
kidney	27	8	5	12	2	73	67	6
liver	28	5	15	1	7	72	70	2
Estradiol-17β 3-benzoate								
rump muscle	90	17	5	59	7	10		
neck muscle	94	20	2	64	6	6		
rump fat	97	29	3	63	2	3		
neck fat	95	23	1	61	6	4		
mesenteric fat	97	35	5	51	4	4		
kidney fat	94	27	4	51	9	7		
kidney	22	6	5	8	4	78	74	4
liver	18	3	5	1	9	82	79	3

Together they account for 80–90% of the total extracted radioactivity. The major metabolites and the qualitative pattern of metabolite occurrence in muscle and fat were the same for animals treated with either estradiol-17β or estradiol-17β 3-benzoate.

Estradiol benzoate was nondetectable in the tissues and fluids analyzed in this study, suggesting rapid hydrolysis of the benzoate ester.

In contrast to muscle and fat, where the metabolites are primarily free estrogens, the major metabolites present in liver and kidney appear in the conjugate fraction. The major conjugate metabolites were the glucuronates, which represented about 85–95% of the total recovered conjugate radioactivity.

CONCLUSION

No major differences were detected in the level or identity of the major forms of estrogen found in the edible tissues of beef animals injected with either [14C]estradiol-17β or [14C]estradiol-17β 3-benzoate. The major form of estrogen in the muscle and fat in both groups of animals was estradiol-17β; estrogen glucuronates were the major forms in liver and kidney.

The highest level of estrogen found in any tissue was only 9.5 ng/g in liver tissue from estradiol benzoate-treated animals, even though these samples were collected at maximum blood levels after extremely high levels of estrogen had been administered.

QUESTIONS AND ANSWERS

T. E. Shellenberger: In your work have you extended your recovery period after last exposure to see when total residue disappeared?

C. C. Kaltenbach: No, we have not done that.

T. E. Shellenberger: Do you have concurrent data on DES?

C. C. Kaltenbach: No.

L. J. Fischer: Were the various fractions (E_1, E_2, E_3) identified only by their elution patterns off the LH column or was an isotope dilution done for each one of those?

C. C. Kaltenbach: For routine purposes, the elution pattern off the column was determined. The methods used have been validated by recrystallization to constant specific activity, micro melting point determinations, and mass spectrometry.

L. J. Fischer: Is it possible to have a 2-hydroxy metabolite in one of the fractions?

C. C. Kaltenbach: Yes, that is possible, but I doubt that it occurred.

D. M. Hendricks: Due to differences in incorporation of radioactivity in various tissues, are the data obtained for high activity tissues (i.e., liver or kidney) more reliable than the data obtained for low activity tissues (muscle)?

C. C. Kaltenbach: Only in terms of statistical validity due to the extent of precision in the samples having higher activity.

D. M. Hendricks: You found no evidence of other metabolites except what you have reported, correct?

C. C. Kaltenbach: No, we have been unable to extract about 30% of the administered radioactivity in liver and kidney tissue. In all other tissues, the total radioactivity could be accounted for by the metabolites.

J. H. Clark: Is benzoate hydrolyzed in the blood?

C. C. Kaltenbach: I do not know as we have not looked for it.

J. H. Clark: Perhaps it could be an esterase in the blood.

C. C. Kaltenbach: Could be.

D. M. Hendricks: What was your rationale for designing the experiment the way you did, that is, to pretreat with estrogens over a period of several days before administering the isotope?

C. C. Kaltenbach: In that these were feedlot cattle, we decided to pretreat the animals so that any estrogen-induced enzyme changes or other effects of the pretreatment would simulate long-time exposure as is encountered in the feedlot situation.

REFERENCES

Carr, B. R., Mikhail, G. and Flickinger G. L. 1971. Column chromatography of steroids on Sephadex LH-20. *J. Clin. Endocrinol. Metab.* 33:358.

Harvey, W. R. 1960. *Least-squares analysis of data with unequal sub-class numbers.* U.S.D.A., ARS-20-8.

Vihko, R. 1966. Gas chromatographic–mass spectrometric studies on solvolyzable steroids in human peripheral plasma. *Acta Endocrinol. Suppl.* 109.

Zaffaroni, A. 1953. Micromethods for the analysis of adrenocortical steroids. *Recent Progr. Hormone Res.* 8:51.

Received June 4, 1975
Accepted June 27, 1975

ESTROGEN CONCENTRATIONS IN BOVINE AND PORCINE TISSUES

D. M. Henricks

Department of Food Science, Clemson University,
Clemson, South Carolina

Using an estrogen antiserum that measures biologically active estrogen (estradiol-17β + estrone), a series of studies have been completed to describe the normal plasma levels of estrogen in cows and sows during various stages of reproduction. The levels are lower in the nonpregnant cow (< 25 pg/ml) than in most species. Only during late pregnancy are the levels easily assayed. The levels in the nonpregnant sow are 2–3 times greater than in the cow. In both species there is a preovulatory peak of estrogen that is responsible for estrus. Plasma estrogen concentrations were also determined after induction of multiple ovulations by exogenous gonadotropin (PMSG) in the cow and sow and after treatment of cows with either a progestin (MGA) or a prostaglandin (PGF$_2$α). (These two drugs have been used to synchronize estrus in the cow.) Estrogen levels rise greatly after treatment with PMSG but show little effect from MGA or PGF$_2$α. Data are also presented on the validation of an assay for E$_2$β and E$_1$ in beef tissue. Less than 20 pg of E$_2$β and E$_1$ was measured per gram bovine muscle tissue. Considerable effort must be expended to apply radioimmunoassay to the assay of estrogens in tissues.

INTRODUCTION

This report summarizes a series of studies that deal in part with: (1) the assay of peripheral plasma levels of "total estrogen" in the untreated cow and sow to ascertain normal levels; (2) the assay of "total estrogen" in cows and sows treated with pregnant mare serum gonadotropin (PMSG), melengesterol acetate (MGA), and prostaglandin F$_2$α (PGF) to ascertain the effects of these agents on estrogen secretion; and (3) the development and validation of an assay and its use to measure estrone (E$_1$) and estradiol-17β (E$_2$β) in edible tissues.

"Total estrogen" levels refer to a value obtained by radioimmunoassay using an antiserum that reacts equally to E$_1$ and E$_2$β. In some of the studies discussed, plasma progesterone and plasma luteinizing hormone (LH) concentrations are presented to demonstrate interrelationships among secretions of the three hormones.

Published with the approval of the Director of the Agricultural Experiment Station as Technical Contribution No. 1282.

Requests for reprints should be sent to D. M. Henricks, Department of Food Science, Clemson University, Clemson, South Carolina 29631.

Journal of Toxicology and Environmental Health, 1:617–639, 1976

RESULTS AND DISCUSSION

Radioimmunoassay for Total Estrogen

The procedure has been described in detail in another article (Henricks et al., 1971). An antiserum prepared against estradiol-17β-succinyl-bovine serum albumin conjugate was generously supplied by B. V. Caldwell. It was used at a dilution of 1:50,000 (initial dilution). At least 5 ml of bovine plasma was allotted per sample to be extracted with freshly opened diethyl ether. To construct a standard curve, a duplicate set of tubes containing 10, 20, 50, 100, and 200 pg of $E_2\beta$ was incubated with antisera. Before adding the estrogen, 5 ml of ether was added to each tube and evaporated. The reproducibility of the standard curve is given in Table 1. The coefficients of variation ranged from 4.1 to 7.0% for the five points on the curve.

Standard curves for $E_2\beta$, E_1, estriol, and $E_2\alpha$ are depicted in Fig. 1. In their ability to displace $[^3H]E_2\beta$ from the antibody, $E_2\beta$ and E_1 were approximately equipotent from 0 to 200 pg/ml estrogen. E_3 and $E_2\alpha$ were much less potent, showing a cross-reaction of less than 8%.

Table 2 presents the method blank and recovery of $E_2\beta$ and $[^3H]E_2\beta$ added to plasma from an ovariectomized cow. Duplicate tubes were run in each experiment. The experiment was replicated 10 times to give a total of 20 values. Values from the samples spiked with $E_2\beta$ were corrected for the method blank. Over the range of estrogen values seen in normally cycling cows, the observed values agreed with the expected values.

Four pools of plasma were each assayed six times to obtain data on the reproducibility of the assay. The results are shown in Table 3. Blood was collected from four cows: from one on day 16, one on day 18 of the estrous cycle, and two on the day of estrus. Using samples of 4 ml, the coefficient of variation ranged from 10 to 21%.

TABLE 1.   Reproducibility of Dose–Response for Estradiol-17β Curve Obtained in 10 Consecutive Assays

Estradiol-17β added (pg)	Percent $[^3H]$estradiol-17β bound to antibody[a] (mean ± SE)	n	cv[b] (%)
0	100		
10	91.4 ± 1.18	10	4.1
20	80.9 ± 1.23	10	4.8
50	61.0 ± 0.89	10	4.6
100	39.8 ± 0.93	9	7.0
200	24.6 ± 0.54	9	6.6

[a]Antibody diluted 1:15,000 (10,000 cpm of $[^3H]$estradiol-17β added per tube).

[b]Coefficient of variation.

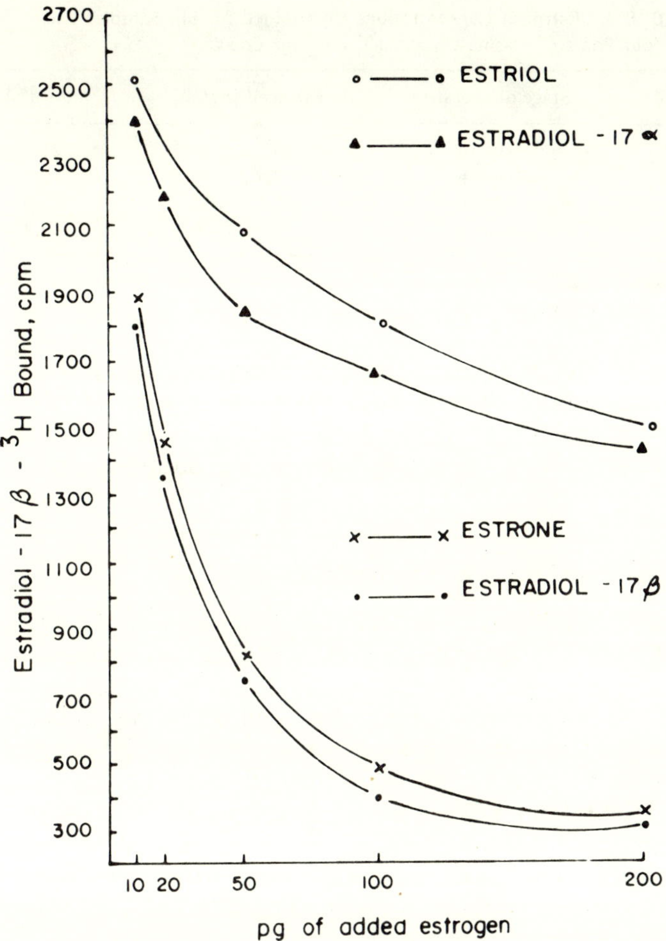

FIGURE 1. Standard dose–response curves for estradiol-17β, estrone, estradiol-17α, and estriol in the radioimmunoassay for plasma estrogens. To each tube was added 100 μl of antibody (dilution 1:15,000) and 136 pg [6,7-^{3}H]estradiol-17β.

TABLE 2. Blank Values and Recovery of Estradiol-17β and [^{3}H]-Estradiol-17β Added to Plasma from an Ovariectomized Cow

	Estradiol-17β (E$_2\beta$) activity (pg) (mean ± SE)	n
4 ml plasma	20.1 ± 0.36	20
4 ml plasma + 50 pg E$_2\beta$	48.6 ± 0.94	20
4 ml plasma + 20 pg E$_2\beta$	18.7 ± 0.55	20
4 ml plasma + 8.5 pg [^{3}H] E$_2\beta$	8.1 ± 0.88	20

TABLE 3. Estrogen Concentrations Determined for Six Samples Each of Four Pools of Plasma Obtained from Four Cows

Pool	Stage of proestrus	Estrogen (pg/ml)[a]	cv[b] (%)
A	Day 16	3.3 ± 0.26	21.0
B	Day 18	10.5 ± 0.44	10.3
C	Estrus	18.8 ± 0.81	10.6
D	Estrus	9.0 ± 0.46	12.8

[a]Mean ± SE.
[b]Coefficient of variation.

Progesterone was assayed in some of these studies by the competitive protein binding assay; diluted dog plasma was used as the source of specific binding protein. In later studies progesterone was determined by radioimmunoassay using antiserum generously supplied by G. D. Niswender.

Plasma Estrogen Levels in Untreated Cows

Peripheral blood plasma was analyzed (Henricks et al., 1972) for total estrogen and progesterone concurrently during several reproductive stages of the cow.

In trial 1, 18 beef heifers were mated, and blood was collected every 3 days until they either returned to estrus or until day 39 of pregnancy. As shown in Fig. 2, in mated cows that returned to estrus, the mean estrogen level was < 7 pg/ml on days 3, 6, 12, and 15 and > 10 pg/ml on day 9 and > 20 pg/ml just before estrus. In mated cows that became pregnant, the mean estrogen level was low (< 5 pg/ml from day 3 until day 39) (Fig. 3). The expected increase in plasma progesterone concentrations until day 12 of the cycle, reflecting a developing corpus luteum (CL), occurred in both the cows returning to estrus and those becoming pregnant. After day 12, the level declined in the nonpregnant cows and increased to 13.9 ng/ml in the pregnant cows. Thus, estrogen concentration is 1000-fold less than progesterone in cows. Its temporal pattern during the cycle is the inverse of the pattern for progesterone, except for the possible blip that occurs at midcycle. The peak that occurs just before estrus is certainly not unexpected if estrogen is required for the behavior accompanying estrus. The basal level of estrogen during early pregnancy is easily rationalized based on the physiological functions of estrogen.

In trial 2, 10 dairy cows were bled daily 7–14 days before parturition, then twice a week for 60 days postpartum. During the 14 days before parturition, estrogen increased from 500 pg/ml to 2,660 pg/ml at parturition (Fig. 4). Plasma progesterone remained relatively steady at one-third the level observed during early pregnancy until one day before parturition, when it fell to 0.7 ng/ml.

After parturition, during the postpartum period, estrogen levels fell precipitously and ranged from nondetectable to 5.6 pg/ml. Prior to an observable estrus, both estrogen and progesterone levels were much lower than those seen after the estrus (Fig. 5).

The estrogen concentrations reported for the estrous cycle, early pregnancy, just before parturition, and the postpartum period are in the same ranges as those that have since been reported by others (Echternkamp and Hansel, 1973; Edquist et al., 1973; Wetteman and Hafs, 1972; Wetteman et al., 1972).

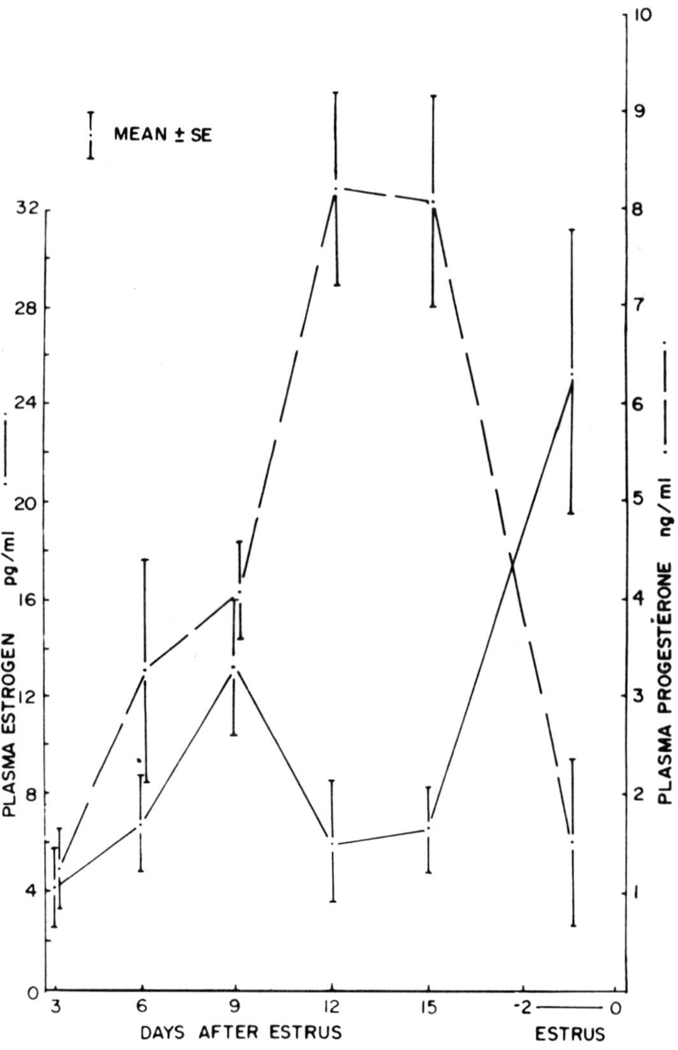

FIGURE 2. Plasma estrogen and progesterone concentrations in mated cows that returned to estrus within the period of an estrous cycle. Mean concentrations ± SE for six cows are depicted.

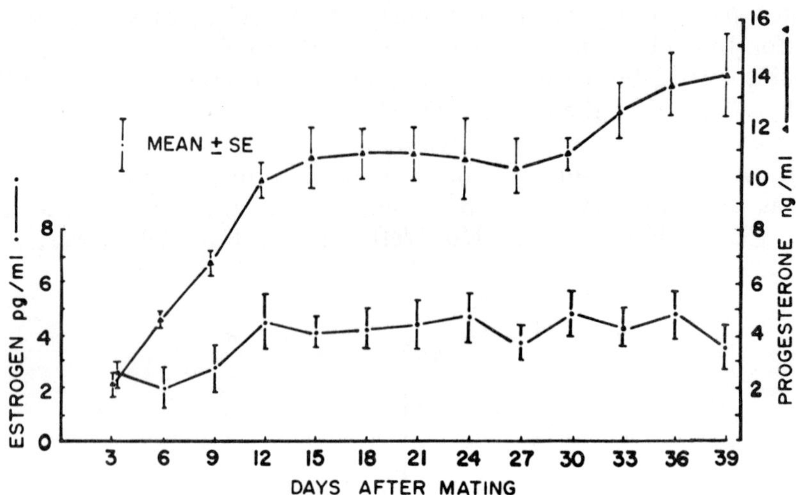

FIGURE 3. Plasma estrogen and progesterone concentrations in mated cows from days 3–39 of pregnancy. Mean concentrations ± SE for 12 cows are depicted.

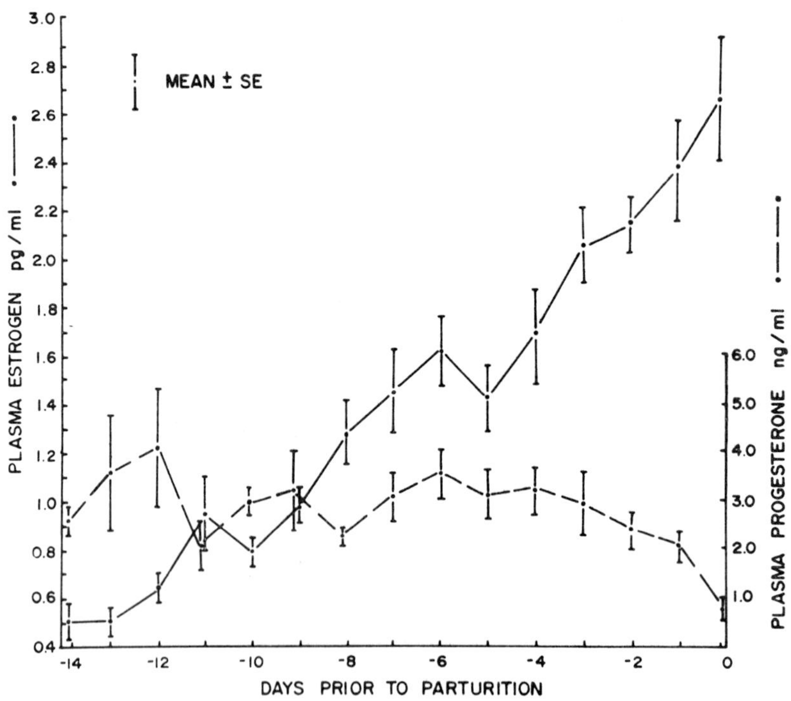

FIGURE 4. Plasma estrogen and progesterone concentrations in cows during the 14-day period prior to parturition. Mean concentrations ± SE for 10 cows are depicted.

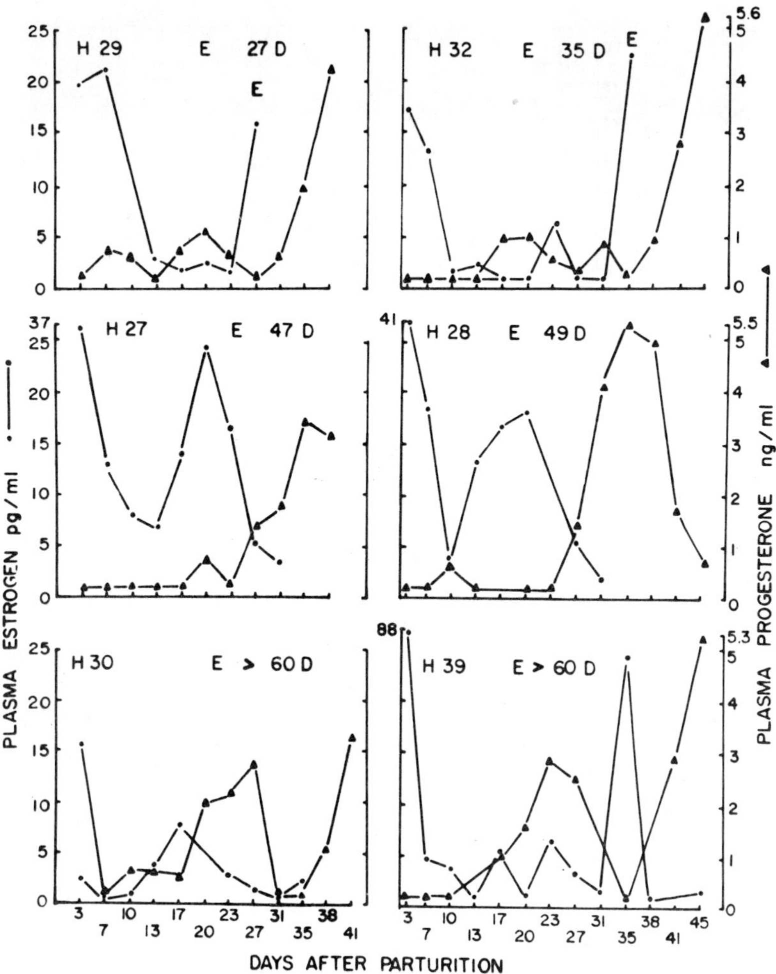

FIGURE 5. Plasma estrogen and progesterone concentrations in individual cows during the postpartum period. Each point is the mean of duplicate determinations. The first postpartum estrus was detected in cows H29 and H32 on days 27 and 35.

Effects of Melengestrol Acetate on Plasma Estrogen Levels in Beef Heifers

MGA, a synthetic progestin, was fed to 25 beef heifers at a rate of 0.5 mg per animal per day for 14 days beginning on day 15 of the estrous cycle. An untreated group of 25 heifers was included in the experiment. The incidence of synchronization, fertilization, and pregnancy rates was determined (Henricks et al., 1973a). Estrogen and progesterone profiles were obtained via intensive bleeding during the MGA feeding period, post-MGA period until estrus, and postmating period.

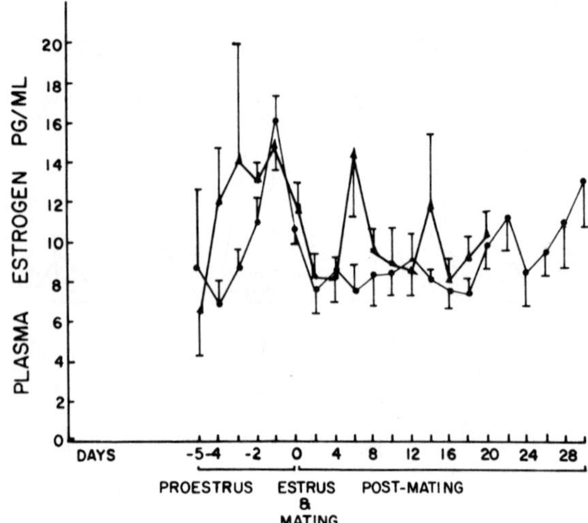

FIGURE 6. Plasma estrogen concentrations in untreated beef heifers. Mean ± SE. Heifers becoming pregnant (●); heifers that returned to estrus (▲).

In general, plasma estrogen levels in the treated heifers were similar to the levels in the untreated heifers, especially during proestrus and after mating (Fig. 6 and 7). This indicates MGA had no lasting effect on the cow's production of estrogen. The estrogen level during the MGA-feeding period was below the level measured just before estrus in both the treated cows and untreated cows. The only difference in estrogen levels were during the proestrous period between animals becoming pregnant and those that did not become pregnant.

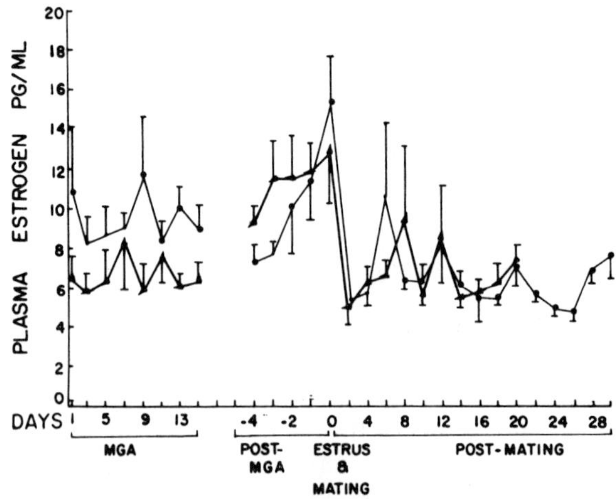

FIGURE 7. Plasma estrogen concentrations in MGA-treated beef heifers in which estrus was synchronized. Mean ± SE. Heifers becoming pregnant (●); heifers that returned to estrus (▲).

Plasma progesterone levels were similar between the treated and untreated groups. In the treated group, however, progesterone levels were less than 1 ng/ml for 13 days before estrus and were in this range for a much shorter proestrous period. It is useful to remember that these animals were treated on day 15, which is near the end of the luteal phase. Of course, estrus is temporarily suspended for as long as MGA is fed. These hormone profiles are similar to those reported in other studies of the effects of MGA (Chow et al., 1972; Dobson et al., 1973).

Effects of PMSG on Plasma Estrogen Levels in Beef Heifers

Doses of 0, 1,600, and 3,200 IU PMSG were administered to groups 1, 2, and 3, respectively, on day 16 of the estrous cycle to 24 beef heifers. Jugular blood was collected twice daily until estrus and at 0, 3, 6, 9, and 12 hr after the beginning of estrus. The blood plasma was analyzed for total estrogen, progesterone, and LH concentrations (Henricks et al., 1973b). The heifers were killed on days 3–8 after mating, and ova and ovaries were examined to determine incidence of fertilization, number of CL, and number of follicles > 10 mm in diameter.

The effect of PMSG treatment on mean plasma estrogen concentrations is shown for the three groups of animals in Fig. 8. The mean rates of increase in estrogen concentration were 0.08, 1.0, and 1.2 pg/ml/hr for groups 1, 2, and 3, respectively, measured over the 4-day period before estrus. The rates were significantly different ($p < 0.05$). Plasma estrogen was not correlated with follicle number at slaughter or the maximum plasma LH level at estrus. For group 3, the maximum estrogen concentration was not correlated with any of these parameters.

The assumed quantity of estrogen secreted during the period of increasing blood levels was related to the number of ovulations. Ovulation rate was generally related to the total amount of estrogen (Fig. 9). During the 12 hr of estrus, estrogen levels in group 2 fell rapidly to a concentration of 10 pg/ml, which is similar to that found in group 1. In group 3 (3,200 IU), however, the level fell only slightly from a peak level of 36 pg/ml to 29 pg/ml. This amount of estrogen may be detrimental to normal physiologic function. Cystic ovaries were found in some of the animals when slaughtered 3 to 8 days later.

PMSG appeared to have a luteotropic effect on progesterone secretion by the CL. Progesterone levels remained elevated for a longer period before estrus and then decreased more rapidly in the animals given 3,200 IU than in the other groups (see Fig. 8). Thus, the "progesterone trough" became of progressively shorter duration before estrus.

The effect of PMSG treatment on plasma LH concentrations during the period before and after estrus is shown in Fig. 10. LH peaks occurred at 0, 3, and 6 hr after the beginning of estrus in cows treated with 3,200, 1,600, or 0 IU PMSG, respectively. Generally, an increase in estrogen is

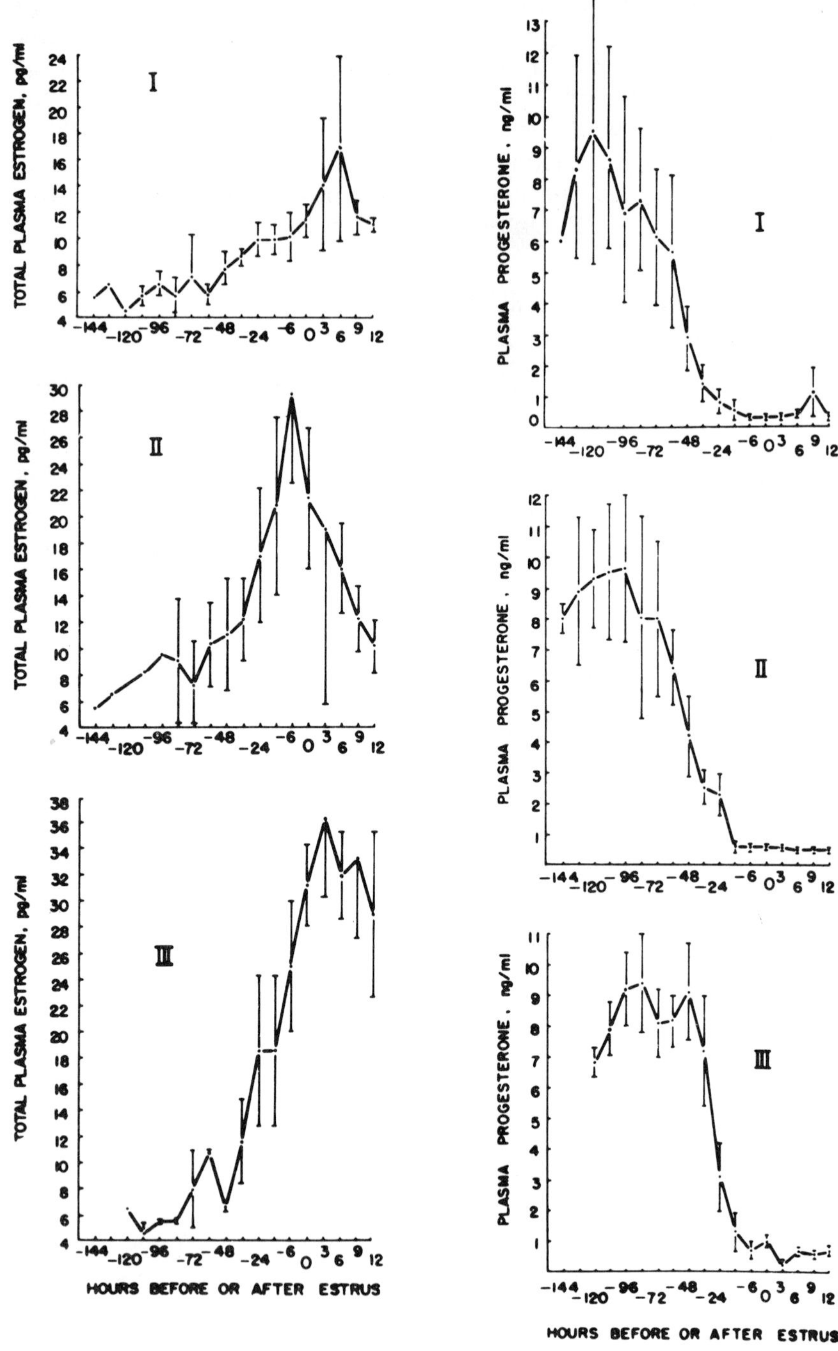

FIGURE 8. Plasma estrogen and progesterone concentrations before and during estrus in beef heifers receiving 0 (I), 1,600 (II), and 3,200 IU (III) PMSG.

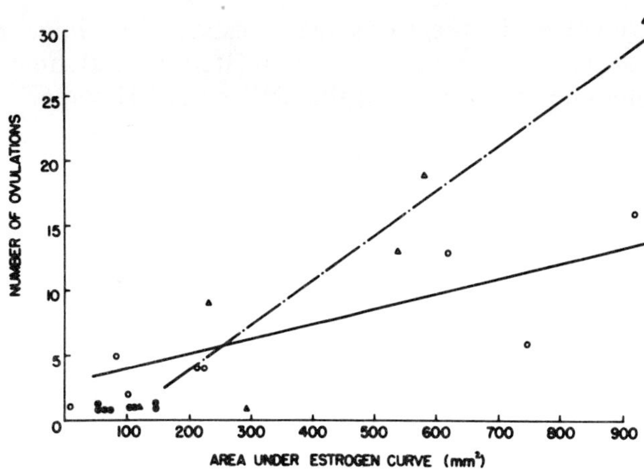

FIGURE 9. A comparison of ovulation rate with the area under the estrogen curve for each cow: 0 IU (●), 1,600 IU (○), 3,200 IU (△).

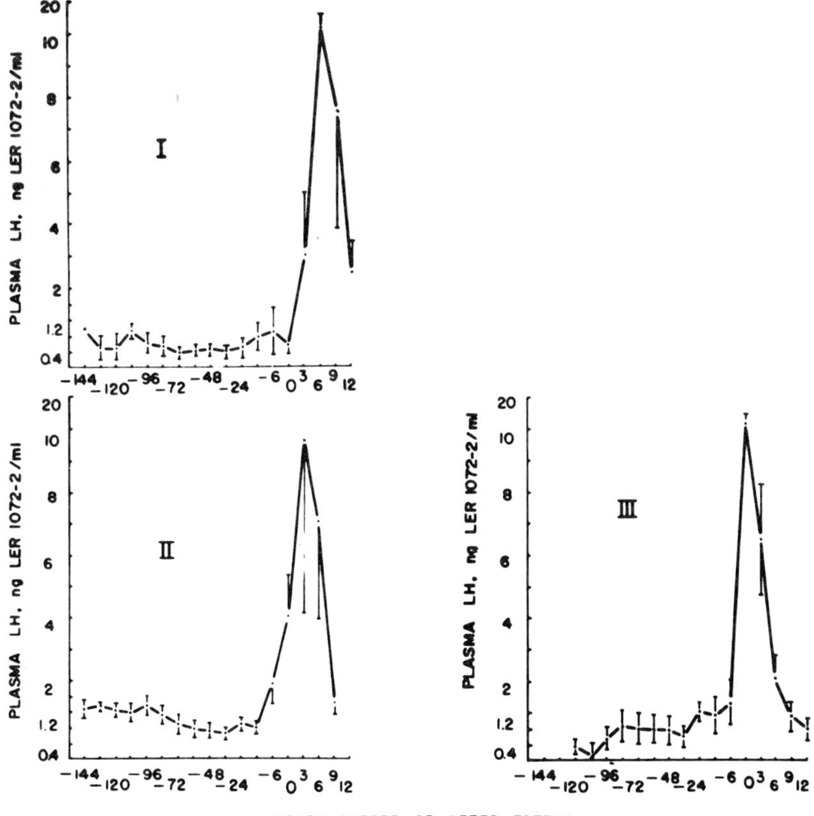

FIGURE 10. Concentrations of LH in plasma before and during estrus in beef heifers receiving 0 (I), 1,600 (II), and 3,200 IU (III) PMSG.

thought to be involved in the LH surge at estrus. Certainly, in this study the quantity of estrogen proved to be related to the time of the LH surge—the greater the estrogen level, the earlier the LH surge.

Effect of PGF$_2\alpha$ on Plasma Estrogen Levels in the Beef Heifer

The effects of prostaglandin F$_2\alpha$ (PGF) on peripheral plasma estrogen and progesterone levels, incidence of premature estrus, and fertility were examined in 30 beef heifers. PGF (2 mg intrauterine) was administered during three periods of the estrous cycle: days 3-5, 9-10, or 16-17. Estrus occurred in 2–5 days following treatment, except in those animals treated on days 3 or 4 of the cycle. As shown in Fig. 11, plasma progesterone levels had fallen greatly by the day following treatment and continued to fall in those heifers returning to estrus. As this occurred, plasma estrogen rose to a peak on the day before or the day of estrus (Fig. 12). Of interest was the fact that estrogen and progesterone patterns were quite similar to those of the untreated animals as they approached estrus, as well as after mating. The main difference between PGF-treated and untreated (Henricks et al., 1974) heifers was the sharper increase in estrogen levels in the treated animals. Similar estrogen levels have been reported by others (W. Hansel, personal communication).

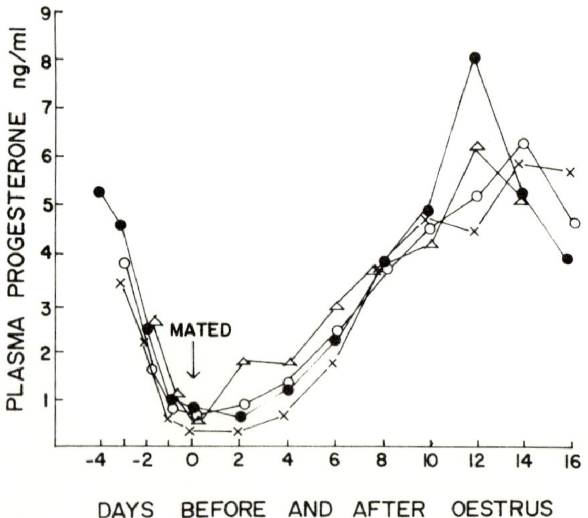

FIGURE 11. Daily plasma progesterone concentrations in untreated heifers (Δ) and after intrauterine administration of PGF$_2\alpha$ on days 3–5 (●), on day 9 or 10 (X), and on day 16 or 17 (○). ↓ = Injection of PGF$_2\alpha$ and estrus.

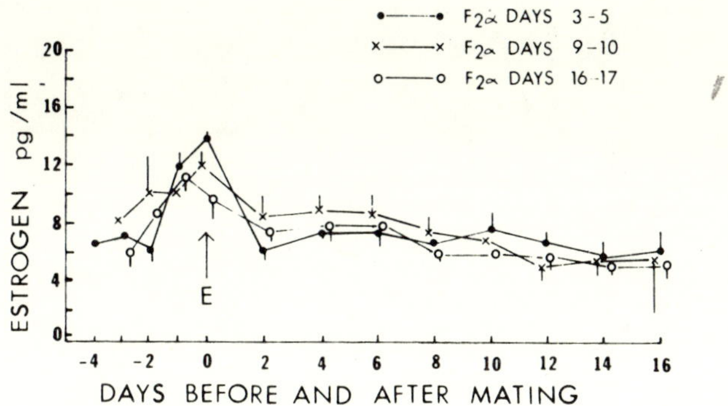

FIGURE 12. Plasma estrogen concentrations in heifers treated with $PGF_2\alpha$ at various times of the estrous cycle.

Effect of PGF and PMSG in Combination on Plasma Estrogen Levels in the Beef Heifer

The design of this experiment is shown in Table 4. Heifers receiving 1,600 IU PMSG exhibited estrogen and progesterone levels similar to those obtained from heifers treated in a previous study (Henricks et al., 1973b). In addition, after mating estrogen levels decreased and remained low, except for a "blip" soon after estrus; conversely, progesterone levels were much higher than in untreated animals (Henricks et al., 1975) (Fig. 13). Heifers receiving the combination treatments had somewhat lower levels of estrogen at estrus and lower levels of progesterone after mating (Figs. 14 and 15). PGF appeared to suppress the ovulatory stimulus caused by PMSG, suggesting a possible means whereby PMSG could be used for multiple births in cattle.

TABLE 4. Experimental Design

Prostaglandin treatment[a] (mg)	Time of PMS treatment[b]	
	2 days before $PGF_2\alpha$	Same day as $PGF_2\alpha$
Control	12[c]	
$PGF_2\alpha$ on[d]		
days 3-5	6	6
days 9-10	6	6
days 16-17	6	6
Total	18	18

[a]$PGF_2\alpha$ given in body of the uterus.
[b]1,600 IU PMS injected subcutaneously.
[c]PMS administered on day 16 of the estrous cycle.
[d]Stage of the unaltered estrous cycle when $PGF_2\alpha$ was given.

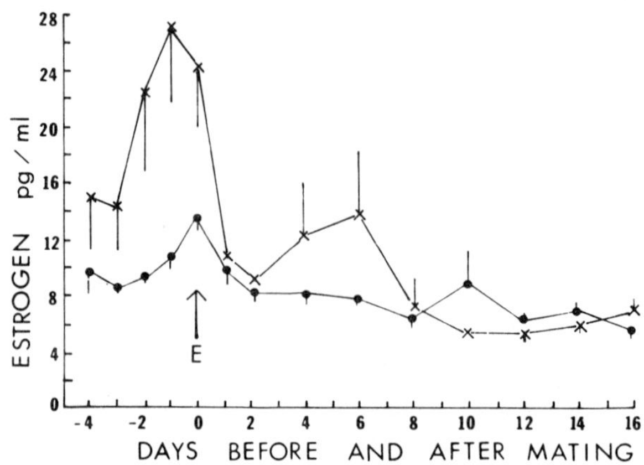

FIGURE 13. Plasma estrogen concentrations in PMS-treated heifers (x) and nontreated heifers (•).

PLASMA ESTROGEN LEVELS IN THE PIG

Proestrus and Early Pregnancy

Plasma estrogen, as well as progesterone and LH, was measured in plasma obtained for the 6 days preceding estrus and on days 3, 6, 12, 18, and 24 of pregnancy. The results are shown in Fig. 16. Plasma estrogen peaked at 40–60 pg/ml at estrus and returned to a baseline of 20–30 pg/ml between periods of estrus, as well as during the first 24 days of pregnancy (Guthrie et at., 1972). This pattern is similar to that of the cow, except that it is 4–5 times greater. Although progesterone levels are also higher in the pig (4–5 times greater), they again show the same pattern

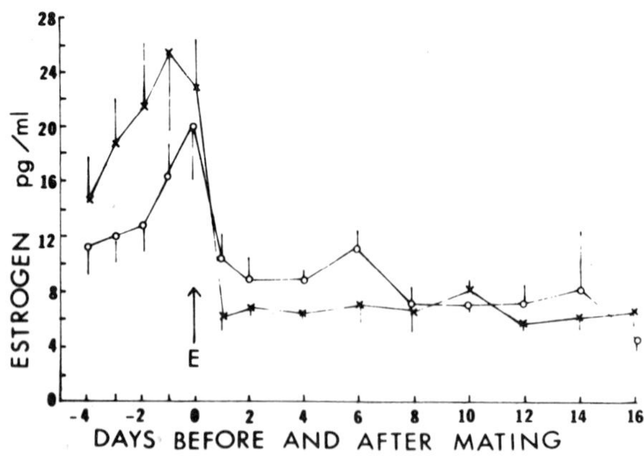

FIGURE 14. Plasma estrogen concentrations in PMSO/PGF$_2\alpha$-treated heifers: PGF 9–10 (x); PGF 16–17 (o).

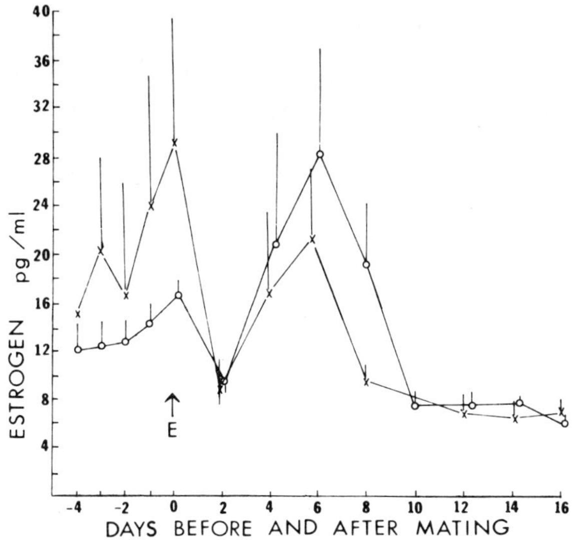

FIGURE 15. Plasma estrogen concentrations in PMS-2/PGF$_2\alpha$-treated heifers: PGF 9–10 (x); PGF 16–17 (o).

except for a fall between days 18 and 24 of pregnancy. This decline is reversed as pregnancy continues. The LH pattern in the pig also resembles that of the cow, except that the concentrations at estrus are markedly lower. The LH peak follows the estrogen peak by 1 or 2 days.

Effect of PMSG on Plasma Estrogen Concentrations

Seventy-one mature gilts were allocated to a 3 X 2 X 2 factorial experiment consisting of three doses of PMSG (0, 600, and 1,200 IU), two levels of food intake, and two times of autopsy after mating (days 6 and 24). Blood plasma was collected from the jugular vein (Guthrie et al., 1975). Plasma estrogen levels before estrus were increased by PMSG treatment as can be seen in Fig. 17. Plasma progesterone levels after mating were also increased by PMSG treatment (Fig. 17). The magnitude of the LH surge on the first day of estrus was not affected by PMSG, but time of estrus and the preovulatory surge was advanced by 1 day. The estrogen levels in the gilts receiving the lesser dose of PMSG (600 IU) more closely resembled the levels in the saline-treated animals than those obtained with the higher dose of PMSG (1,200 IU).

Development of Specific Estrogen Assays for Muscle, Liver, and Kidney Tissues

The application of radioimmunoassay (RIA) to the measurement of estradiol and estrone in these tissues is a task worthy of intensive

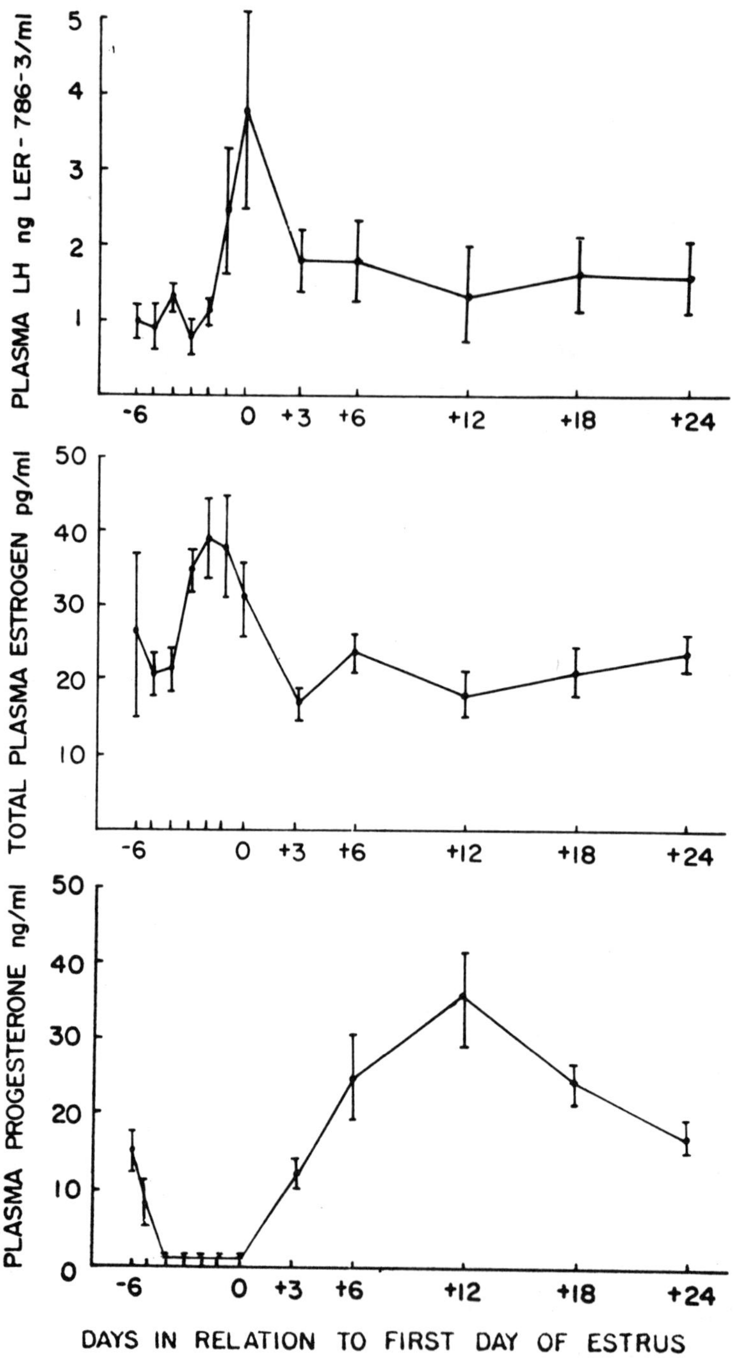

FIGURE 16. Plasma estrogen, progesterone, and LH concentrations before estrus and after mating in gilts.

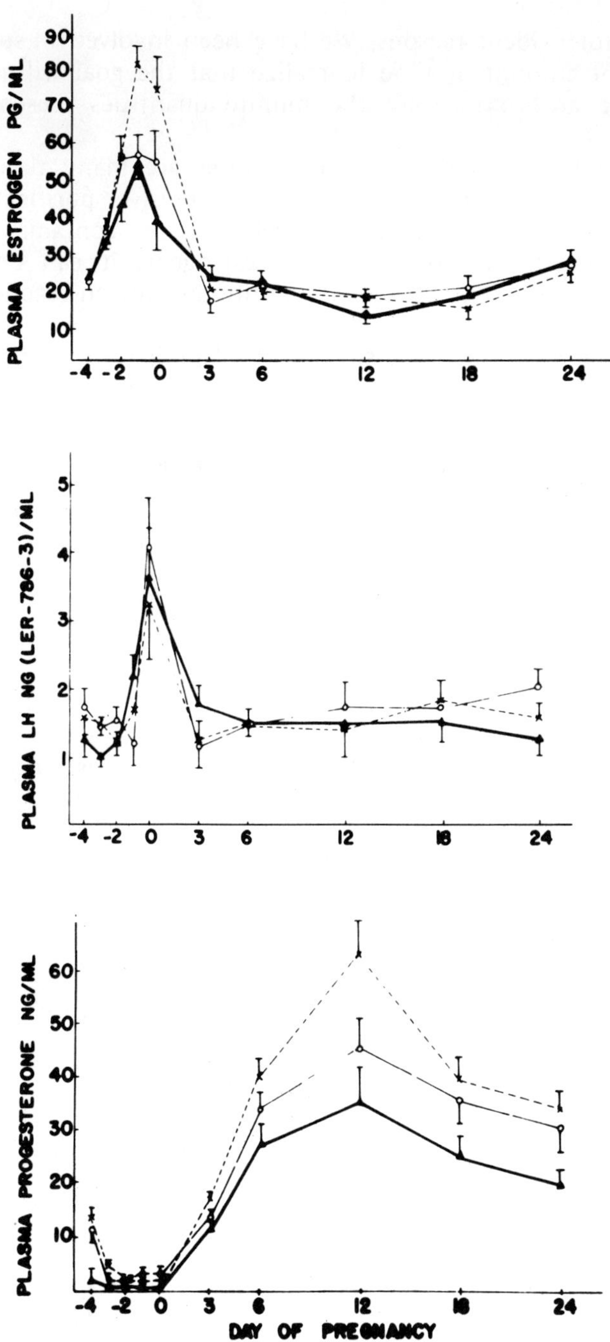

FIGURE 17. Effect of PMSG on plasma estrogen, LH, and progesterone concentrations during proestrus and the first 24 days after mating. Day 0 was the first day of estrus. ▲, no treatment; ○, 600 IU PMSG; x, 1,200 IU PMSG.

investigation for evident reasons. We have been involved in such a project for a sufficient amount of time to realize that the goal will not be easily reached if we are to measure the minute quantities present in muscle tissue.

The contaminants present in these tissues are many times greater in quantity than those present in plasma. Extensive purification mainly through the use of chromatography is needed. If the contaminants that are probably nonspecific remain with the estrogens, it can be shown by scarchard plots that the kinetics of the antibody–antigen reaction is greatly disturbed.

For the purposes of discussion, four "spiking experiments" are included in this report. Table 5 presents the recovery of $E_2\beta$ from an

TABLE 5. Recovery of Estradiol-17β from Tissues by Radioimmunoassay: Expt. 1. Reagent Spiked with E_2 + [^{3}H]E_2

Sample no.	$E_2\beta$ added (pg)	[^{3}H]E_2 recovered (%)	E_2 recovered (pg)	Percent recovery
1	0 (Blank)	61.5	< 20	
2	0	57.9	< 20	
3	0	56.7	< 20	
4	0	47.3	30	
			< 23	
5	90	58.4	98	
6	90	58.9	100	
7	90	62.2	102	
8	90	61.9	90	
			$\bar{X} =$ 98	108.8
9	180	56.9	160	
10	180	53.4	165	
11	180	55.9	135	
12	180	54.8	140	
			$\bar{X} = 150$	83.3
13	360	59.3	292	
14	360	57.2	348	
15	360	51.3	310	
16	360	48.7	330	
			$\bar{X} = 320$	88.8
17	720	55.7	520	
18	720	52.5	640	
19	720	45.3	610	
20	720	50.8	660	
			$\bar{X} = 623$	86.5

TABLE 6. Recovery of Estradiol-17β from Tissues by Radioimmunoassay: Expt. 2. Muscle Spiked with E_2 + [^{3}H]E_2

Sample no.	E_2 added (pg)	[^{3}H]E_2 recovered (%)	E_2 recovered (pg)	
1	0	65.7	20	
2	0	50.7	20	
3	0	65.9	20	
4	0	59.3	20	
5	90	66.9	70	
6	90	47.6	50	
7	90	51.4	70	
8	90	46.4	60	
			$\bar{X} =$ 63	70.0
9	180	40.6	120	
10	180	42.5	120	
11	180	45.3	160	
12	180	53.9	130	
			$\bar{X} =$ 133	73.4
13	360	49.6	170	
14	360	55.8	200	
15	360	66.6	180	
16	360	43.7	270	
			$\bar{X} =$ 205	56.9
17	720	36.8	470	
18	720	41.5	470	
19	720	57.0	590	
20	720	41.3	480	
			$\bar{X} =$ 503	69.9

extraction in which no tissue was included (water not extracted). The recovery of $E_2\beta$ was corrected by the recovery of [^{3}H]$E_2\beta$. In addition to the blank, four levels of $E_2\beta$ were added. Four samples were run per level. The percent recovery ranged from 83 to 109%. Table 6 presents an identical experiment except that muscle was present. Here percent recovery ranged from 57 to 73%. Tables 7 and 8 show a companion experiment in which estrone recovery was determined. From water, percent recovery ranged from 63 to 72%. From muscle, percent recovery of estrone ranged from 78 to 98%.

CONCLUSION

Proestrus, an important phase of the estrous cycle in the cow, has not received the attention it deserves as far as estrogen concentrations are concerned. As reported by Henricks, Dickey, and Hill (1937), mean

TABLE 7. Recovery of Estrone from Tissues by Radioimmunoassay: Expt. 3. Reagent Spiked with $E_1 + [^3H]E_1$

Sample no.	E_1 added (pg)	$[^3H]E_1$ recovered (%)	E_1 recovered (pg)	
1	0	65.0	< 20	
2	0	40.0	30	
3	0	68.4	< 20	
4	0	53.8	< 20	
5	90	63.8	70	
6	90	66.9	60	
7	90	72.2	70	
8	90	64.8	60	
			$\bar{X} = 65$	72.2
9	180	67.5	120	
10	180	58.2	130	
11	180	66.1	110	
12	180	64.0	110	
			$\bar{X} = 118$	65.6
13	360	65.0	240	
14	360	70.5	240	
15	360	61.2	230	
16	360	64.8	200	
			$\bar{X} = 228$	63.3
17	720	64.3	490	
18	720	65.4	460	
19	720	55.7	470	
20	720	62.7	460	
			$\bar{X} = 470$	65.3

estrogen concentrations progressively rise over the 4-day period preceding estrus to reach a peak, ranging from 12 to 25 pg/ml on the day before or day of estrus. The mean estrogen levels as well as mean progesterone levels are presented in Fig. 18. The inverse relationship between the two hormones is apparent. In fact, it is probably a requirement that progesterone levels have begun to fall before estrogen levels can rise significantly. Figure 18 also illustrates the patterns of the two hormones in individual beef heifers.

QUESTIONS AND ANSWERS

S. R. Cernosek: Is the minimum amount detected in your blank (20 pg) based on the standard deviation of your zero dose in your RIA determination?

D. M. Henricks: At present, 10 pg estrone or estradiol-17β differ significantly from the zero dose.

S. R. Cernosek: Since RIA assays are individual determinations, substantial variation must be accounted for in a number of analyses within one or across several experiments. There is some question as to whether the variation should be expressed in terms of standard error or standard deviation.

D. M. Henricks: Either expression is an appropriate means of gauging interassay variation. It is important that the investigator use internal standards and devise some standard of acceptability.

J. H. Clark: The best way to circumvent these various problems in a number of assays is to adjust the concentration of all reagents to that portion of the standard curve that gives a linear response.

D. M. Henricks: Unlike plasma, tissues such as muscle, liver, and fat often change the nature of the standard curve to such an extent as to

TABLE 8.   Recovery of Estrone from Tissues by Radioimmunoassay: Expt. 4. Muscle Spiked with $E_1 + [^3H]E_1$

Sample no.	E_1 added (pg)	$[^3H]E_1$ recovered (%)	E_1 recovered (pg)	
1	0	85.3	< 20	
2	0	64.9	< 20	
3	0	76.6	< 20	
4	0	62.9	30	
5	90	68.0	70	
6	90	48.5	90	
7	90	49.7	100	
8	90	53.7	90	
			$\bar{X} = 88$	98.0
9	180	47.6	170	
10	180	49.5	140	
11	180	62.5	150	
12	180	49.1	150	
			$\bar{X} = 152$	84.4
13	360	39.0	290	
14	360	44.6	350	
15	360	55.8	250	
16	360	59.3	$\bar{X} = 240$	
			283	78.6
17	720	40.2	520	
18	720	42.5	600	
19	720	57.1	630	
20	720	41.9	520	
			$\bar{X} = 563$	

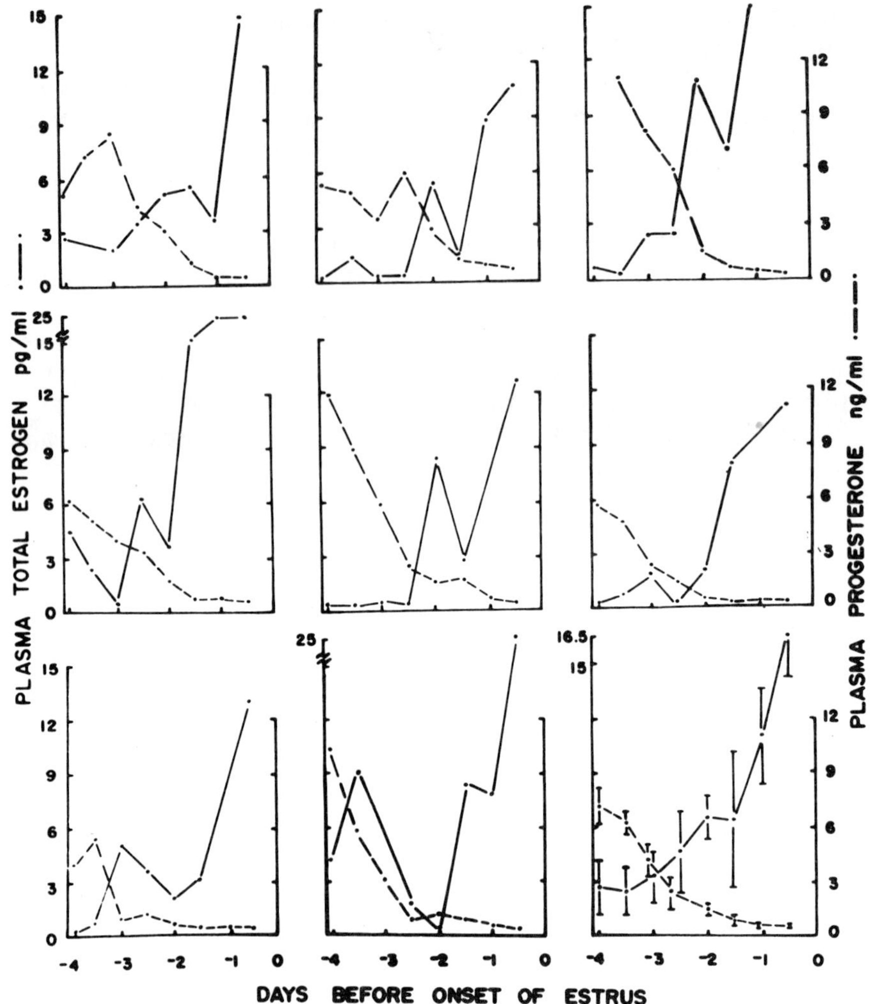

FIGURE 18. Concurrent changes in plasma estrogen and progesterone concentrations during proestrus in the beef heifer.

render it of questionable value. To minimize such problems, one must eliminate as many contaminants as possible from tissues before conducting RIA analysis. Nonpolar steroids, especially progesterone and testosterone, are difficult to clean up for RIA analysis.

C. C. Kaltenbach: Often blanks are not linear over a range, and subtracting a common blank often renders RIA data obtained in such tissue of doubtful value.

J. H. Clark: The total blood level of estrogen during proestrus is about 6×10^{-11} *M* and if 10% of that is free for interaction, the level of active estrogen is 10^{-12} *M*. At these extremely low levels estrogen would not interact with receptors or other tissues; in other words, it would not work in a cow or even in a test tube. Either the method indicated here is greatly underestimating the amount of estrogen (which I doubt), or something else is going on, of which we know very little. The 10^{-11} M value is 100-fold lower than the dissociation constant for estrogen receptors.

D. M. Henricks: I think it is also very important to realize that peripheral blood hormone levels may be quite different from hormone levels at the site of the target tissues.

REFERENCES

Chow, L. A., Thatches, W. W., Chenault, J. C., Kalva, P. S. and Wilcox, C. J. 1972. Effects of MGA on bovine plasma ovarian steroids. *J. Anim. Sci.* (64th Annu. Mtg.) 35:239. (Abstr.)

Dobson, H., Boyd, L. J. and Exley, D. 1973. Circulating estrogens and progesterone in the bovine after synchronization with MGA. *J. Reprod. Fertil.* 33:231–237.

Echternkamp, S. E. and Hansel. W. 1973. Concurrent changes in bovine plasma hormone levels prior to and during the first postpartum estrous cycle. *J. Anim. Sci.* 27:1362–1370.

Edquist, L. E., Ekman, L., Gustafsson, B. and Johansson, E. D. B. 1973. Peripheral plasma levels of oestrogens and progesterone during late bovine pregnancy. *Acta Endocrinol.* 73:81–88.

Guthrie, H. D., Henricks, D. M. and Hanklin, D. L. 1972. Plasma estrogen, progesterone and luteinizing hormone prior to estrus and during early pregnancy in pigs. *Endocrinology* 91:675–679.

Guthrie, H. D., Henricks, D. M. and Handlin, D. L. 1975. Plasma hormone levels and fertility in pigs induced to superovulate with PMSG. *J. Reprod. Fertil.* 41:361–370.

Henricks, D. M., Dickey, J. F. and Hill, J. R. 1971. Plasma estrogen and progesterone levels in cows prior to and during estrus. *Endocrinology* 89:1350–1355.

Henricks, D. M., Dickey, J. F., Hill, J. R. and Johnson, W. A. 1972. Plasma estrogen and progesterone levels after mating and during late pregnancy and postpartum in cows. *Endocrinology* 90:1336–1341.

Henricks, D. M., Hill, J. R. and Dickey, J. F. 1973. Plasma hormone levels and fertility in beef heifers treated with melengestrol acetate (MGA). *J. Anim. Sci.* 37:1169–1175.

Henricks, D. M., Hill, J. R., Dickey, J. F. and Lamond, D. R. 1973b. Plasma hormone levels in beef heifers with induced multiple ovulations. *J. Reprod. Fertil.* 35:225–233.

Henricks, D. M., Long, J. T., Hill, J. R. and Dickey, J. F. 1974. The effect of prostaglandin $F_2\alpha$ during various stages of the estrous cycle of beef heifers. *J. Reprod. Fertil.* 41:113–120.

Henricks, D. M., Dickey, J. F., Long, J. T. and Hill, J. R. 1975. Effects of PMS and $PGF_2\alpha$ on gonadal hormones and reproduction in the beef heifer. *J. Reprod. Fertil.* 35:225–233.

Wettemann, R. P. and Hafs, H. D. 1972. Gonadal and pituitary hormone after insemination. *J. Anim. Sci.* (65th Annu. Mtg.) 35:257. (Abstr.)

Wettemann, R. P., Hafs, H. D., Edgerton, L. A. and Swanson, L. V. 1972. Estradiol and progesterone in blood serum during the bovine estrous cycle. *J. Anim. Sci.* 34:1020–1024.

Received June 4, 1975
Accepted June 27, 1975

STUDIES ON METABOLISM AND EFFECTS OF ESTROGEN ON PITUITARY PROLACTIN AND LH SECRETION

C. L. Chen, M. L. Pattison, L. R. Engleking,
R. R. Gronwall

Department of Physiological Sciences, College of Veterinary
Medicine, Kansas State University, Manhattan, Kansas

The effect of a subcutaneous injection of estradiol on the secretion of pituitary prolactin in the rat and the relationship between serum estradiol level and luteinizing hormone (LH) secretion in mare were reviewed. In addition, the effect of estradiol injection on LH secretion and the metabolism of [^{14}C]estradiol in intact and bile duct fistulated pony mares were studied. Low (0.1 µg/day/rat) to moderate dose (5 µg/day/rat) of estradiol benzoate injected subcutaneously to mature or immature rats significantly increased pituitary content of prolactin and serum prolactin level five- to tenfold. On the other hand, high dose of estradiol (10 µg/day/rat or more) was less effective in stimulating prolactin secretion, and it appeared that progesterone injected concurrently with estradiol had some inhibitory action on the stimulatory effect of estradiol. Studies in pony mares showed that the physiologic level of serum estradiol during proestrus was important for the induction of the ovulatory surge of LH. Intramuscular injection of a low dose (2 or 4 mg/mare) of estradiol was stimulatory, whereas a high dose (8 mg/mare) was inhibitory for LH secretion in pony mares. Results of the estradiol metabolism studies indicated a relatively long half-life for estradiol in the mare. The majority of the [^{14}C]estradiol metabolites appeared in the urine within 24 hr following intravenous injection. Enterohepatic circulation appeared to be important for estradiol metabolism in mare.

INTRODUCTION

Considerable work has appeared on the effect of estrogen on the secretion of pituitary prolactin and LH (Chen and Meites, 1970; Nagasawa et al., 1969; Tsai and Yeu, 1971; Voogt et al., 1970; Yamaji et al., 1971). Most *in vitro* and *in vivo* studies indicate that estrogen stimulates prolactin synthesis and release. Furthermore, estrogen is one of the most important factors in promoting mammary gland growth and enhancing DMBA-induced mammary tumor development. The evidence demonstrated the importance of estrogen in the regulation of pituitary prolactin secretion and mammary development. The importance of estrogen in the regulation of pituitary LH secretion has been stressed by others (Tasi and Yeu, 1971; Yamaji et al., 1971). Many laboratories have reported the metabolism of various synthetic estrogenic compounds in a large number of mammalian species including

This work was supported in part by KSU Agriculture Experimental Station and NIH grant AM11384 to Dr. Gronwall.

Requests for reprints should be sent to C. L. Chen, Department of Physiological Sciences, College of Veterinary Medicine, Kansas State University, Manhattan, Kansas 66505.

Journal of Toxicology and Environmental Health, 1:641–655, 1976

man (Fotherby and James, 1972). However, many of the metabolism studies were performed in the laboratory animals. This presentation attempts to review some of the work on the interrelationship among estrogen, prolactin, and LH secretion in rats and mares. In addition, research on the estrogen metabolism and its effect on mare LH secretion are discussed.

RESULT AND DISCUSSION

Effect of Estradiol on Pituitary Prolactin Secretion

Determination of the minute amounts of serum prolactin levels was made possible by the development of radioimmunoassay (Niswender et al., 1969). This radioimmunoassay system, determined to be specific for rat prolactin, did not cross-react with other rat pituitary hormones. However, the rabbit anti-rat prolactin antiserum appeared to cross-react slightly with hamster and gerbil pituitary prolactin (Fig. 1). Numerous reports on the influence of various agents on the pituitary prolactin secretion have appeared since the development of the rat prolactin radioimmunoassay (Meites et al., 1972). This presentation will be limited to the discussion on the influence of estrogen on prolactin secretion.

A number of investigators have suggested that low or moderate doses of estrogen stimulate, whereas high doses inhibit pituitary prolactin secretion (Tolley and Malpress, 1948). Inhibition, however, has never been demonstrated. Our study shows that serum prolactin concentration is markedly increased by estradiol administration, and even high doses of estradiol do not inhibit prolactin release (Fig. 2) (Chen and Meites, 1970). Progesterone has relatively little ability to increase serum prolactin values but can partially counteract the stimulatory effects of small doses of estradiol (Chen and Meites, 1970). Similar results are observed in the

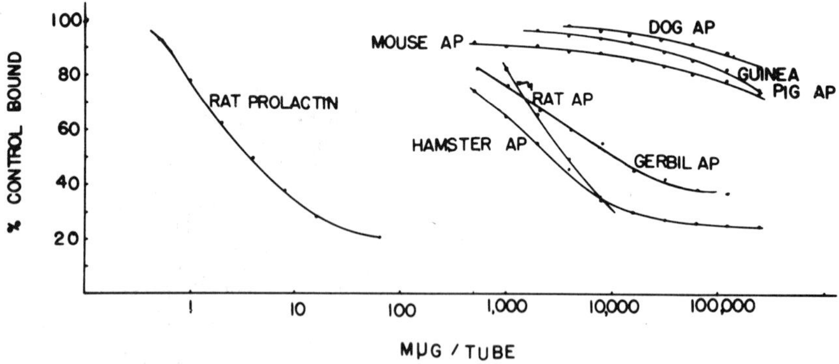

FIGURE 1. Standard curves for radioimmunoassay of prolactin from pituitary preparations from different species. Rat prolactin, purified rat pituitary prolactin; AP, pituitary homogenate.

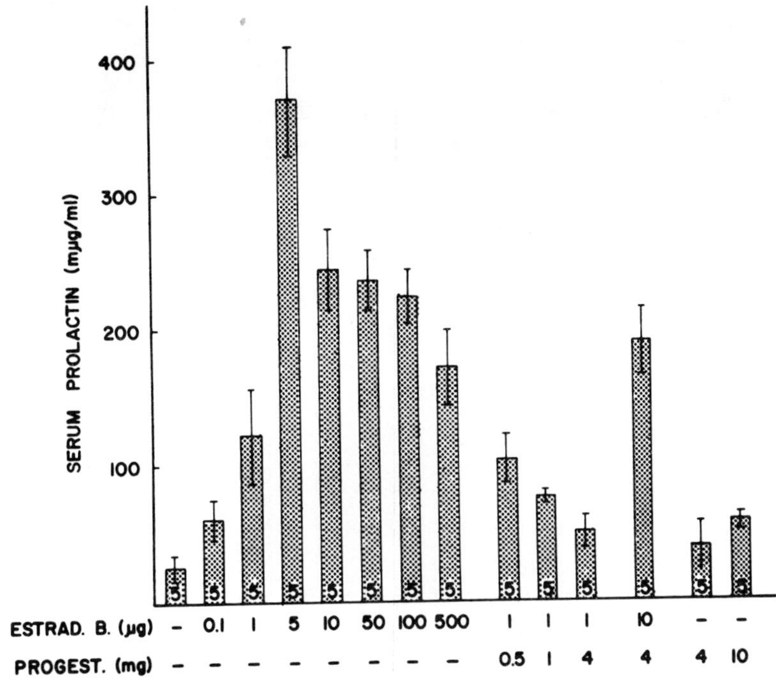

FIGURE 2. Effects of different doses of estradiol benzoate and/or progesterone on serum prolactin concentration in ovariectomized rats (five rats/group). The vertical bars indicate standard error of the mean. From Chen and Meites (1970).

pituitary prolactin concentration following estradiol and progesterone administration (Fig. 3). These results show that the inhibitory action of postpartum lactation in women by estrogen is exerted directly on the mammary gland, not by reducing pituitary prolactin secretion.

Administration of estradiol benzoate to 30-day-old female rats exhibited similar effects in increasing pituitary prolactin secretion (Fig. 4) (Voogt et al., 1970). Large doses of estradiol benzoate appeared to be less effective in stimulating prolactin synthesis and release. In addition to these direct stimulatory effects of estradiol on the pituitary prolactin secretion, the action of estradiol also appeared to stimulate prolactin release through reducing hypothalmic prolactin inhibiting factor. Implantation of estradiol into the median eminence markedly increases the prolactin release (Fig. 5) (Nagasawa et al., 1969). Implantation of estrogen in other areas of the brain or implantation of cholesterol in the median eminence has no effect on the pituitary prolactin secretion. Therefore, estrogen may exert its stimulatory action on the hypothalamus, as well as on the pituitary gland, to enhance the synthesis and release of prolactin.

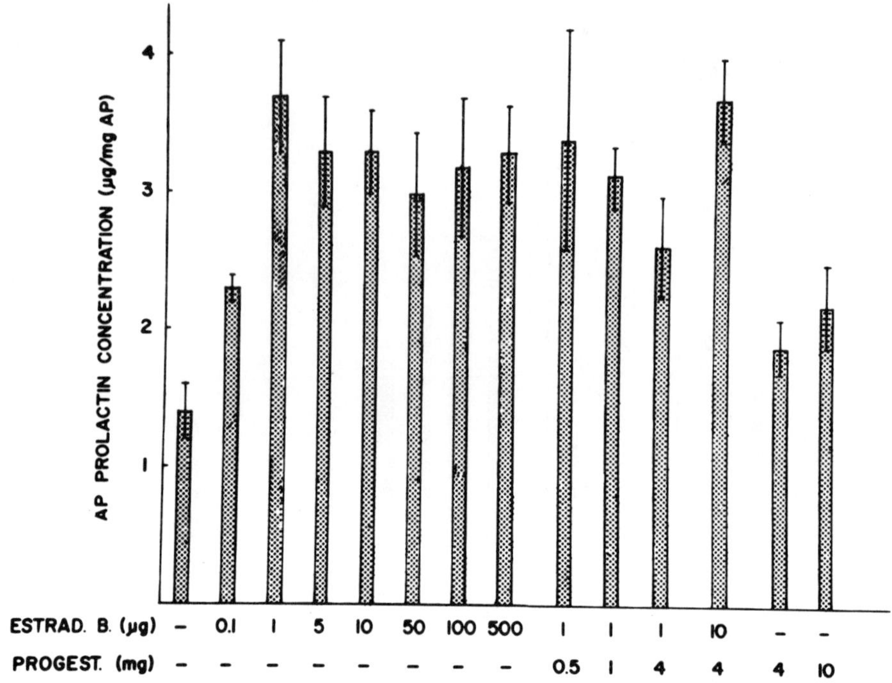

FIGURE 3. Effects of different doses of estradiol benzoate and/or progesterone on anterior pituitary prolactin concentration in ovariectomized rats (five rats/group). The vertical bars indicate standard error of the mean. From Chen and Meites (1970).

Effect of Estrogen on the Pituitary LH Secretion

It has long been recognized that the tonic LH secretion is under a negative feedback control of the gonadal hormones. Fichera (1905) was the first to show that an interruption of the negative feedback loop by gonadectomy in several different species and in both sexes was followed by enlargement of the pituitary. Burrows (1949) and Van Reese (1964) reported that the removal of the gonads in either sex leads to an enhanced output of gonadotropins. Ovariectomy of the rhesus monkey results in an increase of plasma LH 2 days postoperative and reaches a plateau (10 times basal concentration) 20 days after the operation (Atkinson et al., 1970). On the other hand, an increase of peripheral circulating levels of estrogen during a discrete period of the reproductive cycle of man (Tsai and Yeu, 1971), rat (Brown-Grant et al., 1970; Caligaris et al., 1971; Ferin et al., 1969; Labhestwar, 1972), sheep (Goding et al., 1969), and monkey (Yamaji et al., 1971) may be the primary stimulus for the preovulatory LH surge. Estradiol administration sustained longer than 12 hr in ovariectomized monkey (Yamaji et al., 1971) resulted in an LH surge identical to spontaneously occurring LH surges.

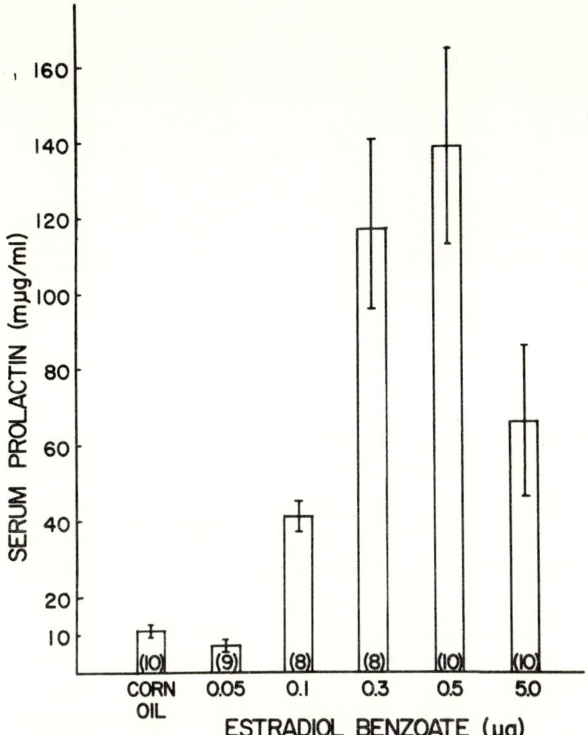

FIGURE 4. Serum prolactin levels of 30-day-old female rats following estradiol benzoate administration. The vertical bars at the top of each column indicate SEM. Numbers on bottom of each column indicate total number rats per group. After Voogt, Chen, and Meites (1970).

In our study mares were used as experimental animals to study the interrelationship between serum levels of estradiol and LH secretion. A radioimmunoassay of ovine LH, adopted for determining equine LH (Fig. 6) (Pattison et al., 1974), was shown to be specific for LH (Niswender et al., 1968). Serum estradiol levels were determined by radioimmunoassay method of Odell et al. (1969). The antiserum was specific for estrogen but not for estradiol (Fig. 7). Therefore, after the steroids were extracted with ether, the estradiol fraction was obtained by chromatograph using Sephadex LH-20 eluted with benzene:methanol mixture (85:15). Only the estradiol fraction was collected and measured by radioimmunoassay (Fig. 8).

In normal estrous cycles (eight estrous cycles of six mares) all mares had long duration of increased serum LH and estradiol during estrus (Fig. 9). In all instances serum estradiol levels reached a peak 1–2 days prior to LH peak. The summarized results in Fig. 10 clearly demonstrate that serum estradiol level reached a peak approximately 2 days before LH peak. These results demonstrate how gradual increases in estradiol level at this reproductive period serve as stimulus for the LH surge.

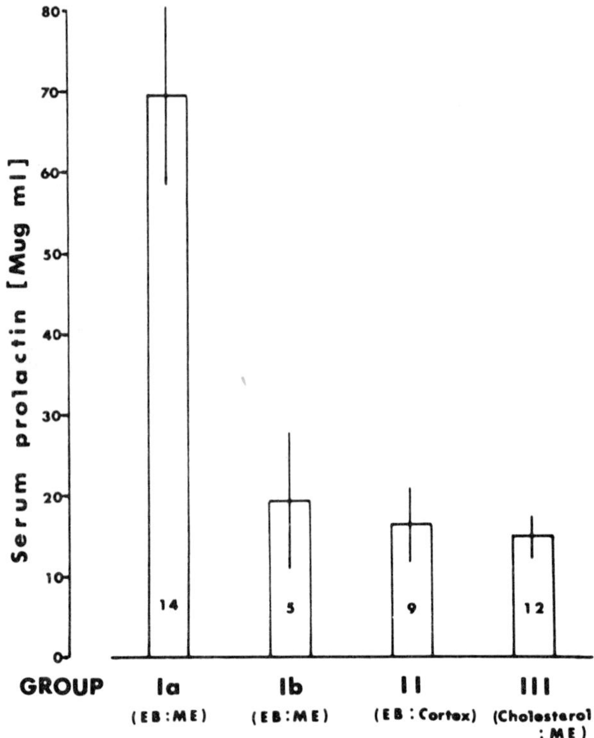

FIGURE 5. Effects of estradiol benzoate (EB) or cholesterol implant in median eminence (ME) or cerebral cortex on serum prolactin levels of rats with mammary tumors. Center vertical bars indicate SEM. Number of rats is indicated at the bottom of each column. Group Ib represents rats with improper estradiol benzoate implant outside of the median eminence. From Nagasawa, Chen, and Meites (1969).

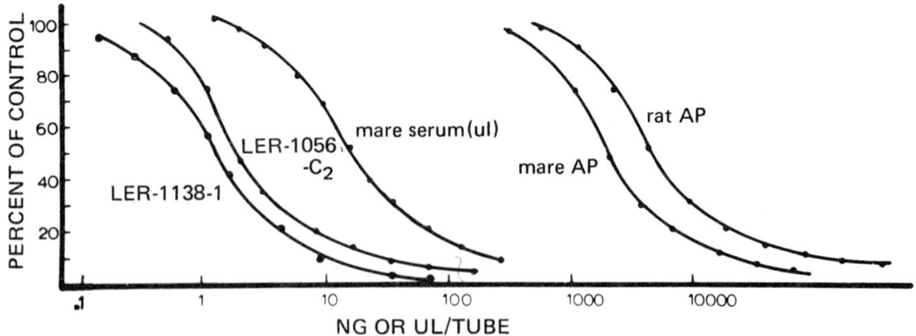

FIGURE 6. Dose–response curves in an ovine radioimmunoassay system for ovine LH (LER-1056-C$_2$), equine LH (LER-1138-1), mare pituitary extract, mare serum from one mare at ovulation, and rat pituitary extract. Each point represents the mean of triplicate assays. From Pattison, Chen, Kelley, and Brandt (1974).

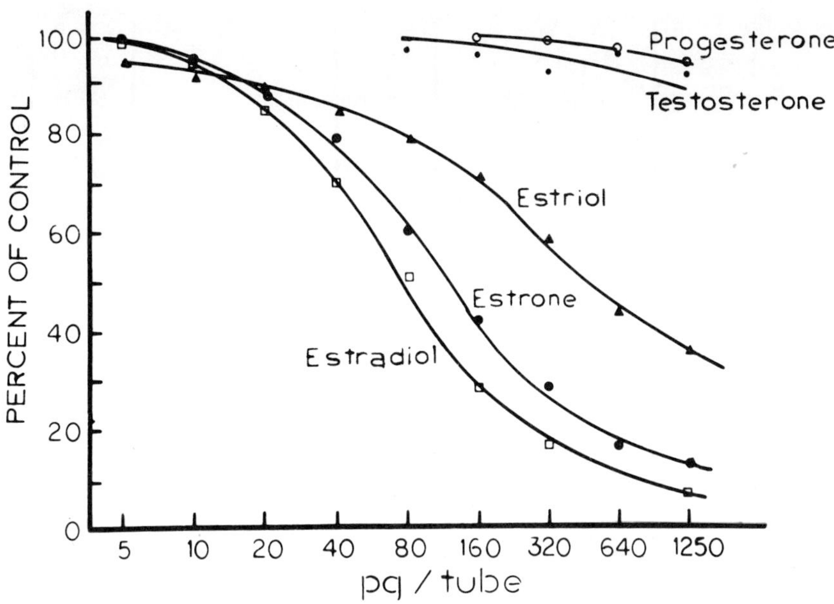

FIGURE 7. Dose–response curves for estradiol, estrone, and estriol using radioimmunoassay. Each point represents the mean triplicate assays. From Pattison, Chen, Kelley, and Brandt (1974).

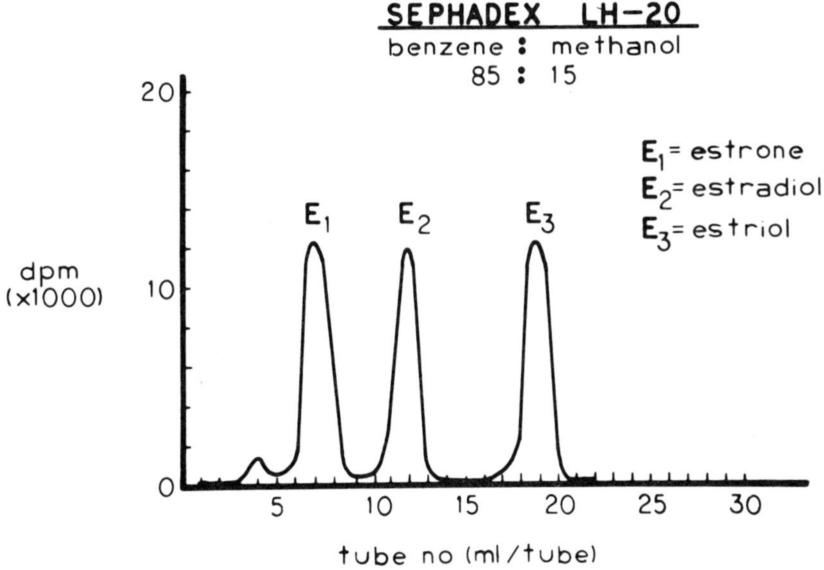

FIGURE 8. Elution pattern of radioactive estrogens in Sephadex column (1 X 20 cm). Benzene:methanol (85:15) mixture was used as eluant.

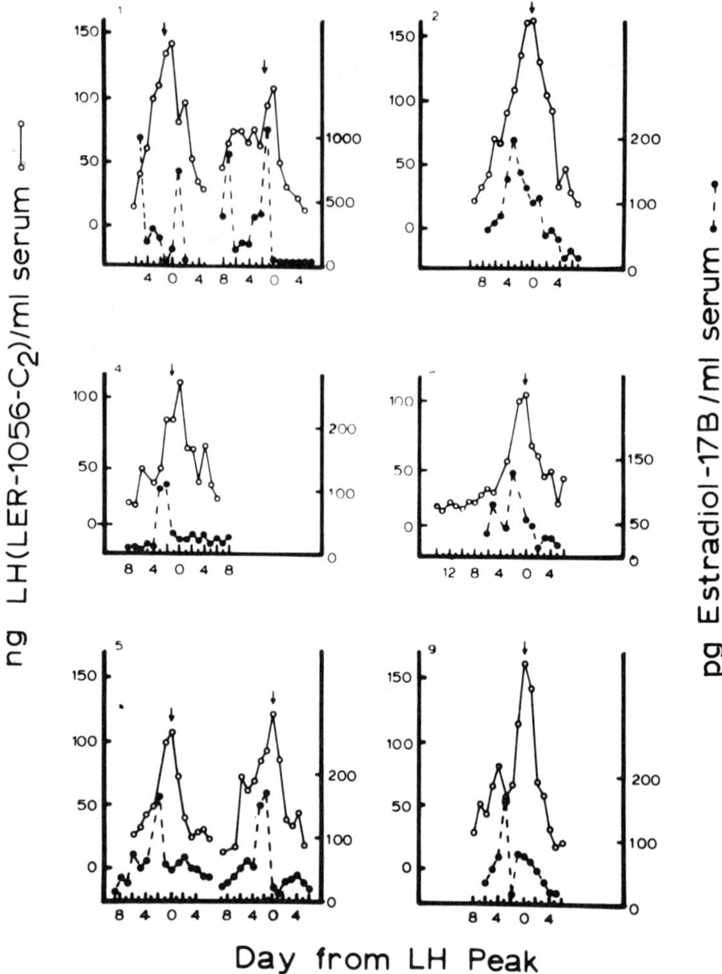

FIGURE 9. Serum levels of LH and estradiol during eight estrous cycles of six mares. Day 0 is the day of the LH peak. Arrow indicates the day of ovulation determined by rectal palpation. From Pattison, Chen, Kelly, and Brandt (1974).

It appears that the secretion pattern of equine reproductive hormones is rather unique. In contrast to the short surge of LH during the peak behavioral estrus and short time prior to ovulation in most other mammalian species (Gay et al., 1970), high serum levels of LH in the mare were maintained 5–10 days, coinciding with the long duration of estrus in the mares. Ovulation occurs on the day of, or in some instances the day after, the LH peak. Perhaps the long duration of high serum estradiol and LH is due to the relative longer half-life for equine LH (Parlow, 1961a,b). The half-life of estradiol in the mare will be discussed below.

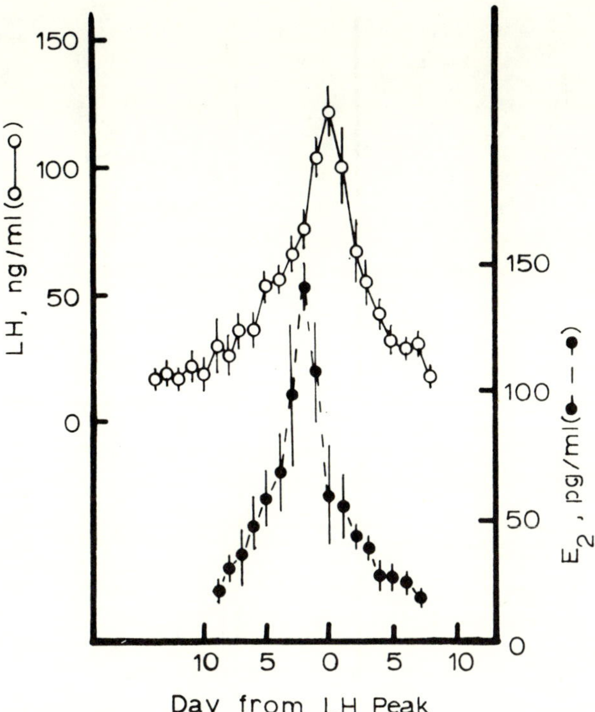

FIGURE 10. Composites of the mean levels of LH and estradiol of estrous cycles normalized to the day of LH peaks. The vertical bars represent the standard error of the mean. From Pattison, Chen, Kelley, and Brandt (1974).

To further evaluate the relationship of serum estradiol levels and pituitary LH secretion, a single dose of either estradiol or estradiol benzoate was injected intramuscularly to diestrous or ovariectomized pony mares. Figure 11 illustrates the effects of 2–8 mg estradiol or estradiol benzoate on the serum levels of estradiol and LH. Administration of 2–4 mg of either estradiol or estradiol benzoate increased serum estradiol levels within 2 hr following injection. Serum estradiol levels were increased to estrous levels (100 pg/ml) 4 hr following injection. Serum LH levels showed an initial decrease then increased for approximately 16 hr after injection of estrogen. These results showed that 2 or 4 mg estradiol administered intramuscularly can increase serum estradiol levels to that of physiologic estrous levels, thus resulting in a surge of LH. This is in good agreement with the results reported in other species (Tsai and Yeu, 1971; Yamaji et al., 1971).

On the other hand, intramuscular injection of 8 mg estradiol benzoate to a diestrous pony mare failed to induce any increase in serum LH levels, though serum estradiol levels were increased several fold over the estrous levels. Possibly, the high serum estradiol level exerted a negative feedback on the pituitary gland.

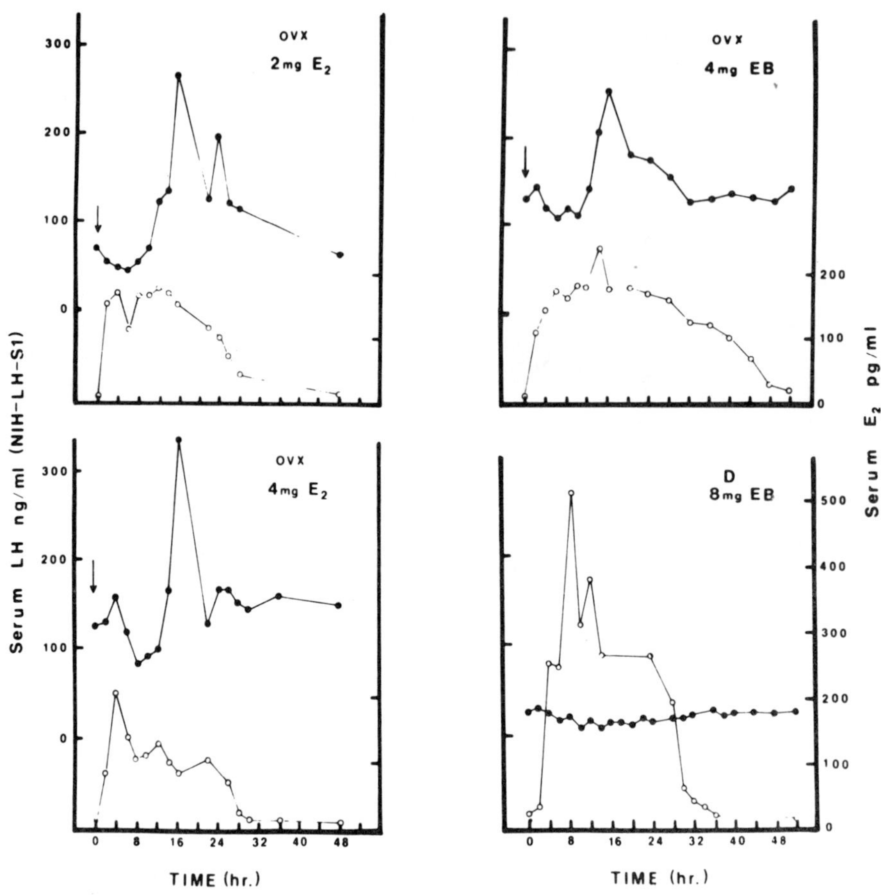

FIGURE 11. Effects of a single dose of estradiol (E_2) or estradiol benzoate (EB) on the serum levels of estradiol and LH in ovariectomized (OVX) or diestrous (D) mares. The arrows indicate the time of intramuscular injection of the hormone.

Metabolism of [^{14}C] Estradiol in the Pony Mares

Disappearance of [^{14}C] estradiol in the plasma and the excretion of the radioactivity in the urine were studied in two diestrous pony mares. The excretion of radioactivity in bile was also studied in one of the two mares with bile fistula. Ten μCi [^{14}C] estradiol (New England Nuclear, specific activity 50 mCi/mmol) was injected into the jugular vein of each mare. Blood samples were collected through an indwelling cannula in the other jugular vein at various time intervals for 24 hr. Urine samples were collected via ureter catheter at 1-hr intervals. Bile samples were collected via the bile fistula and the bile flow rates determined.

The disappearance of [^{14}C] estradiol in the plasma of two mares is summarized in Fig. 12. It appeared that [^{14}C] estradiol was rapidly bound as

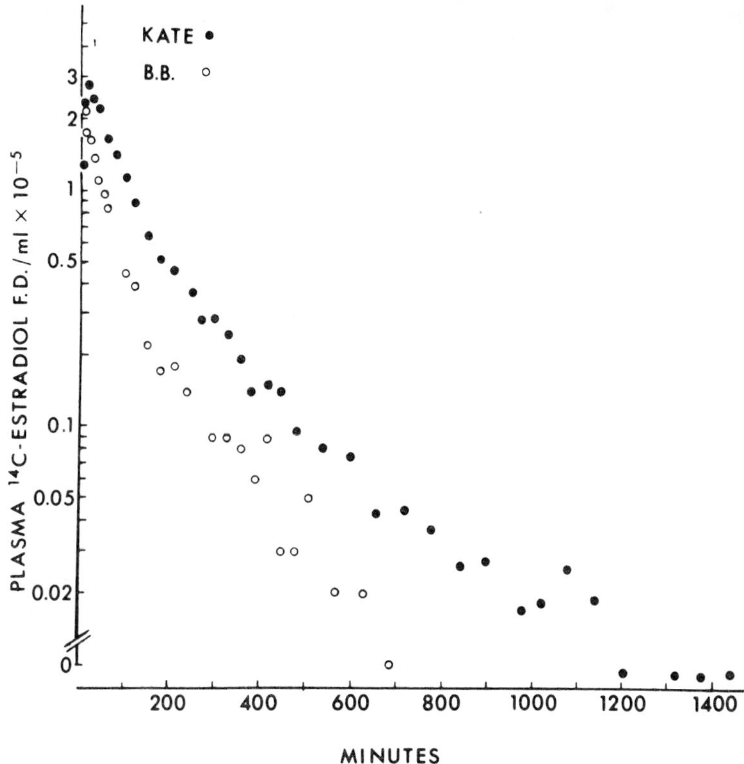

FIGURE 12. Plasma disappearance curves for [^{14}C]estradiol following intravenous injection of 10 μCi [^{14}C]estradiol in two mares. The radioactivity is expressed in fractional dose (F.D.)/ml plasma.

soon as it was injected and gradually reappeared in the circulation. The half-life of the radioactivity in the plasma was approximately 40 min for one mare and 70 min for the other. The half-life of the second part of the curve appeared to be 2–3 hr. These results appear to indicate that the half-life of estradiol may be longer in the mare than most other species reported (Fotherby and James, 1972).

The majority of the radioactivity was excreted in the urine within 4–5 hr following intravenous injections (Fig. 13); peak radioactivity occurred at 1–2 hr. Radioactivity can be detected in the urine throughout the 24 hr collection period. Figure 14 depicts the accumulative dose of [^{14}C]-estradiol in the urine for 24 hr. The accumulative radioactivity was 55 and 65%, respectively, for these two pony mares at the end of 24 hr. Others have reported that between 23 and 47% of radioactivity was recovered from the urine in 5 days from human subjects who received synthetic estrogen (Reed et al., 1970). Beach et al. (1967) and Layne and Williams (1967)

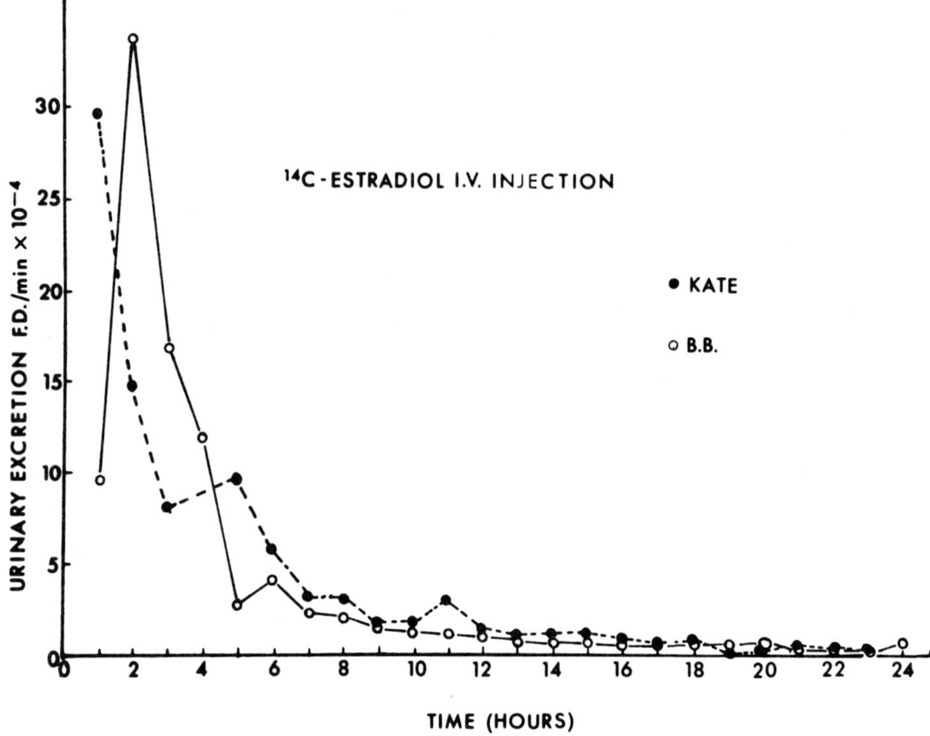

FIGURE 13. Urine excretion of radioactivity following intravenous injection of 10 μCi [^{14}C] estradiol in two mares. The radioactivity is expressed in fractional dose (F.D.)/min for 24 hr.

reported 54% of [^{14}C] synthetic estrogen in the urine of rabbits over a 5-day period.

Figure 15 shows the radioactivity in the bile of one pony mare. Radioactivity appeared within 1 hr following intravenous injection and reached a peak in 1.5 hr. The radioactivity then declined to a low by 3 hr. The radioactivity in the bile indicated a recycle of [^{14}C] estradiol took place, pointing out the possible importance of enterohepatic circulation of estradiol. A lower but relatively steady radioactivity was found over a 24-hr period.

CONCLUSION

Low to moderate doses of estrogen appear to stimulate the secretion of prolactin and LH in most mammalian species, while high doses of estrogen prove less effective in inducing prolactin secretion and inhibiting the secretion of LH in the mares. Results of the metabolism studies indicate that estradiol may have a longer half-life in the mare and that most metabolites appear in the urine within 24 hr. Enterohepatic circulation of estradiol was shown to be important in the mare.

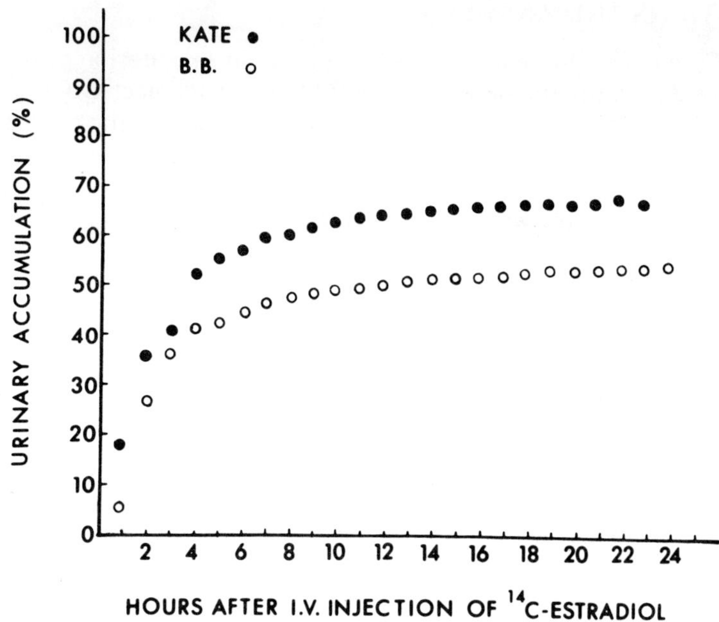

FIGURE 14. Accumulative amount of radioactivity excreted in the urine for 24 hr in two mares. Approximately half the radioactivity is excreted within the first 4 hr. The total recovery of the radioactivity was 55 and 65%, respectively.

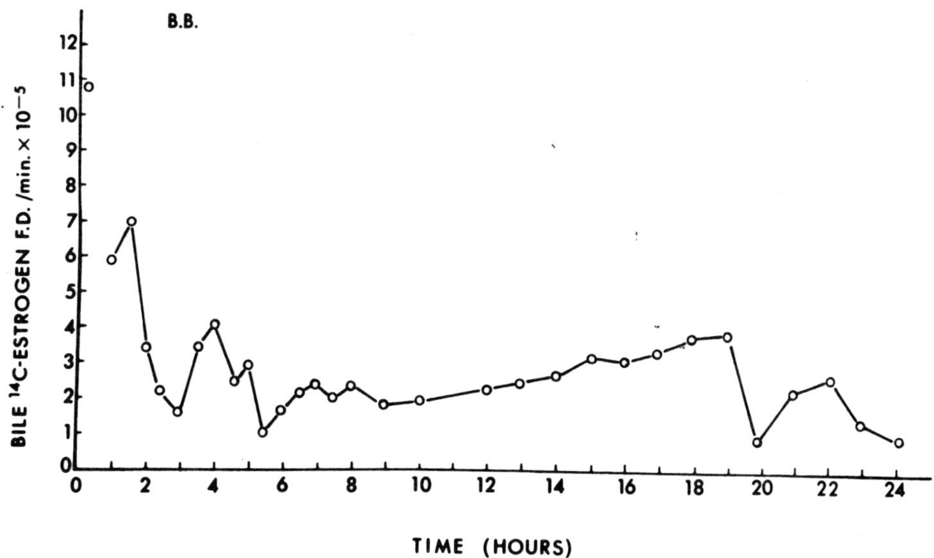

FIGURE 15. Biliary excretion of radioactivity following intravenous injection of [^{14}C]estradiol in one mare with bile fistula. The bile radioactivity is expressed in fractional dose (F.D.)/min for 24 hr.

QUESTIONS AND ANSWERS

S. R. Cernosek: Did you use mouse prolactin or anterior pituitary as a source of prolactin for the data you presented on cross reactivity?

C. L. Chen: The rat anterior pituitary was a source of the prolactin used in these studies.

Unidentified: What was your source of LH antibodies?

C. L. Chen: I used ovine LH antibodies obtained from G. D. Niswender at Colorado State.

REFERENCES

Atkinson, L. E., Bhattacharya, A. N., Monroe, S. E., Dierschke, D. J. and Knobil, E. 1970. Effects of gonadectomy on plasma LH concentration in the rhesus monkey. *Endrocrinology* 87:847–849.

Beach, V. L., Steinetz, B. G., Giannina, T. and Melia A. 1967. Fate of orally administered quinestrol and ethyinl estradiol in rabbits. *Int. J. Fertil.* 12:148–154.

Brown-Grant, K. D., Naftolin, E. and Naftolin, F. 1970. Peripheral plasma oestradiol and luteinizing hormone concentrations during the oestrus cycle of the rat. *J. Endocrinol.* 48:295–296.

Burrows, H. 1949. *Biological actions of sex hormones,* 2d. ed., pp. 40–82. London and New York: Cambridge Univ. Press.

Caligaris, L., Astrada, J. J. and Taleisnik, S. 1971. Release of luteinizing hormone induced by estrogen injection into ovariectomized rats. *Endocrinology* 88:810–815.

Chen, C. L. and Meites, J. 1970. Effects of estrogen and progesterone on serum and pituitary prolactin levels in ovariectomized rats. *Endocrinology* 86:503–505.

Ferin, M., Tempone, A., Zimmering, P. E. and Vande Wille, R. L. 1969. Effect of antibodies to 17-β-estradiol and progesterone on the estrous cycle of the rat. *Endocrinology* 85:1070–1078.

Fichera, B. 1905. Sur l'hypertrophe di la glande pituitaire consecutive a la castration. *Arch Ital. Biol.* 43:405–426.

Fotherby, K. and James, F. 1972. Metabolism of synthetic steroids. *Adv. Steroid Biochem. Pharmacol.* 3:67.

Gay, V. C., Midgley, A. R., Jr. and Niswender, G. D. 1970. Patterns of gonadotropin secretion associated with ovulation. *Fed. Proc.* 29:1880–1887.

Goding, J. R., Catt, K. J., Brown, J. M., Kaltenback, C. C., Cumming, I. A. and Mole, B. J. 1969. Radioimmunoassay for ovine luteinizing hormone. Secretion of luteinizing hormone. Secretion of luteinizing hormone during estrous and following estrogen administration in the sheep. *Endocrinology* 85:133–142.

Labhestwar, A. P. 1972. Role of estrogens in spontaneous ovulations: Evidence for positive feedback in hamsters. *Endocrinology* 90:941–946.

Layne, D. S. and Williams, K. I. H. 1967. Urinary metabolites of radioactive quinestrol in rabbits. *Int. J. Fertil.* 12:158–163.

Meites, J., Lu, K. H., Wuttke, W., Welsch, C. W., Nagasawa, H. and Quadri, S. K. 1972. Recent studies on functions and control of prolactin secretion in rats. *Recent Progr. Hormone Res.* 28:471–526.

Nagasawa, H., Chen, C. L. and Meites, J. 1969. Effects of estrogen implant in median eminence on serum and pituitary prolactin levels in the rat. *Proc. Soc. Exp. Biol. Med.* 132:859–861.

Niswender, G. D., Roche, J. F., Foster, D. L. and Midgley, A. R., Jr. 1968. Radioimmunoassay of serum levels of luteinizing hormone during the cycle and early pregnancy in ewes. *Proc. Soc. Exp. Biol. Med.* 129:901–904.

Niswender, G. D., Chen, C. L., Midgley, A. R., Meites, J. and Ellis, S. 1969. Radioimmunoassay for rat prolactin. *Proc. Soc. Exp. Biol. Med.* 130:793–797.

Odell, W. D., Abraham, G., Randar, H., Swerdloff, R. S. and Fisher, D. A. 1969. Influence of immunization procedures on titer, affinity and specificity of antisera to glycoproteins. In *Immunoassay of gonadotropins,* ed. E. Diczfalusy, vol. 1, pp. 54–76. Stockholm: Karolinska Institutet.

Parlow, A. F. 1961a. A new parameter for biological characterization of gonadotropins: Species differences and significances for bioassay. *Proc. 43d Mtg. Endocrine Soc.*, p. 3. (Abstr.)

Parlow, A. F. 1961b. Bio-assay of pituitary luteinizing hormone by depletion of ovarian ascorbic acid. In *Human pituitary gonadotropin*, ed. A. Albert, pp. 300–312. Springfield, Ill.: Charles C Thomas.

Pattison, M. L., Chen, C. L., Kelley, S. T. and Brandt, G. W. 1974. Luteinizing hormone and estradiol in peripheral blood of mares during estrous cycle. *Biol. Reprod.* 11:245–250.

Reed, M. J., Kamyab, S., Steele, S. J. and Fotherby, K. 1970. *Excerpta Med. Int. Congr. Ser.* No. 210, p. 171.

Tolley, S. J. and Malpress, F. H. 1948. Hormonal control of mammary growth. In *The hormones*, ed. G. Pincus and K. V. Thimann, pp. 695–744. New York: Academic Press.

Tsai, C. C. and Yeu, S. S. 1971. Acute effects of intravenous infusion of 17β-estradiol on gonadotropin: Release in pre- and post-menopausal women. *J. Clin. Endocrinol. Metab.* 32:766–771.

Van Reese, G. P. 1964. Interplay between steroid sex hormones and secretion of FSH and ISCH. In *Major problems in neuroendocrinology*, ed. E. Bajusz and G. Jasmin, pp. 322–345. Basel: Karger.

Voogt, J. L., Chen, C. L. and Meites, J. 1970. Serum and pituitary prolactin levels before, during and after puberty in female rats. *Am. J. Physiol.* 218:396–399.

Yamaji, T., Dierochke, D. J., Hotchkiss, J., Bhattacharyer, A. N., Surve, A. H. and Knobil, E. 1971. Estrogen induction of LH release in rhesus monkey. *Endocrinology* 89:1034–1041.

Received June 4, 1975
Accepted June 27, 1975

REPRODUCTIVE ENDOCRINOLOGY OF FEMALE CHIMPANZEES: A SUITABLE MODEL OF HUMANS

William Hobson, Frederick Coulston

The International Center of Environmental Safety, Albany Medical College, Holloman Air Force Base, New Mexico

Charles Faiman, Jerremy S. D. Winter, Francisco Reyes

Endocrine-Metabolic Sections of the Health Science Center, Winnipeg, and Department of Physiology, Pediatrics, Obstetrics and Gynecology, University of Manitoba, Canada

Similarities between reproductive processes in humans and chimpanzees have led to speculation that the chimpanzee might be an excellent reproductive-endocrine model of humans. Data comparing patterns and concentrations of serum gonadotropins, prolactin, and sex steroids in female humans, chimpanzees, and rhesus monkeys sustain this concept. The striking evidence that levels of estrone, estradiol, estriol, progesterone, and chorionic gonadotropin are similar during human and chimpanzee pregnancy support the contention that chimpanzees, like humans but unlike rhesus monkeys, have a definitive fetoplacental unit. Likewise, during normal menstrual cycles serum patterns of LH, FSH, progesterone, and estradiol are similar in women and chimpanzees, but differ in rhesus monkeys. Thus by using the chimpanzee it may be possible to accurately assess the safety for human use of compounds that might affect reproductive-endocrine processes without direct exposure to humans.

INTRODUCTION

Recent controversy surrounding safety evaluation of hormonally active compounds used or consumed by humans underscores the need for an appropriate animal model from which hormonal effects may be predicted for humans. Although certain aspects of reproductive endocrinology of rodents are similar to those of humans, significant differences between the reproductive endocrinology of humans and rodents preclude their use as a final model for humans. These difficulties have led to an increasing use of nonhuman primates as a human endocrine model. Their use, principally of the rhesus monkey, has led to greater understanding of human endocrinology; however, as data became available, significant differences in the reproductive endocrinology of humans and rhesus monkeys have become

This work was supported in part by the Federal Department of Research and Technology of the Federal Republic of Germany, under a coordinated research program entitled "Ecologic-Toxicologic Effects of Foreign Compounds in Non-Human Primates and Other Laboratory Animals," and in part by Medical Research Council of Canada, grant MT 2997.

Requests for reprints should be sent to William Hobson, International Center for Environmental Safety, P.O. Box 1027, Holloman AFB, New Mexico 88330.

Journal of Toxicology and Environmental Health, 1:657–668, 1976

apparent. For instance, metabolism of the estrogens and progesterone is dissimilar in these two species, and differences in both the patterns and concentrations of serum steroid hormones during pregnancy are especially striking.

These differences led to interest in other species as possible endocrine models of humans and, especially, to the search for an appropriate model of human pregnancy. Recent data have suggested that the chimpanzee may be an appropriate surrogate for humans in reproductive endocrinology studies. This communication reviews some of the important similarities of chimpanzee and human steroid metabolism and presents data demonstrating the similarity of human and chimpanzee reproductive endocrine patterns during menstrual cycles and pregnancy.

RESULTS AND DISCUSSION

Reproductive Patterns

Female chimpanzees undergo menarche at about 8 yr of age (Keeling and Roberts, 1972). At full sexual maturity chimpanzees have regular menstrual cycles of 28–36 days' duration. Unlike the rhesus monkey, they cycle during the entire year. In our colony the mean length of gestation is 229 days, a figure similar to previous reports (Keeling and Roberts, 1972). An interesting aspect of the chimpanzee menstrual cycle is the marked swelling of the perineal area during the preovulatory phase of the menstrual cycle. This provides an additional visible means of following reproductive cyclicity. The swelling has been shown to be induced by estrogen and can be blocked by administration of progesterone (Graham et al., 1972). Menopause has not been documented in chimpanzees; however, they have been observed to reproduce until age 35 (Guilloud, 1969). Thus, the temporal aspects of chimpanzee reproduction are like those of humans.

Animal Handling

In the past the use of chimpanzees for studies requiring frequent blood sample collection has been discouraged by the apparent difficulty of handling these animals. We have found, however, that in a stable colony they can be trained, by using rewards of fruit, to voluntarily submit to frequent blood collections. Although it is possible to train adult animals in this manner, it is less difficult if the initial training is accomplished before puberty.

Urinary Hormone Patterns

Patterns of urinary steroid excretion in chimpanzees appear to be qualitatively similar to those in humans; however, total amounts of steroids excreted are less. As in humans, the principal urinary metabolite of progesterone in chimpanzees is pregnanediol (Elmadjian and Forchielli,

1965; Guilloud, 1969), which is mainly present in the glucuronide form (Guilloud, 1969). The pattern of pregnanediol excretion corresponds to that of the human, but the amounts present in the urine are below the human range (Guilloud, 1969). In contrast, progesterone is thought to be mainly excreted as androsterone (Jeffersy, 1966; Plant et al., 1969) in rhesus monkeys.

In chimpanzees, rhesus monkeys, and baboons, estrone is the major estrogen present in the urine during the menstrual cycle (Guilloud, 1969; Hopper and Tullner, 1970; Stevens et al., 1970), whereas in humans estriol is secreted in equal or greater amounts (Brown and Carey, 1963). In humans and chimpanzees the urinary peaks of estrone during the follicular and luteal phases of the cycle are approximately equal. This luteal estrone peak is lower in the rhesus monkey and absent in the baboon (Hopper and Tullner, 1970; Stevens et al., 1970). Following injection of [^{14}C]estrone, Jirku and Layne (1965) identified labeled estradiol-17β, estriol, and 2-methoxyestrone in the urine of a pregnant chimpanzee. 2-Methoxyestrone has not been identified as a urinary estrogen metabolite in any other species except man and thus appears to be unique to the higher primate species.

Serum Hormone Levels

Radioimmunoassays. In contrast to experience with rhesus monkeys (Faiman et al., 1967; Neill et al., 1967b) human radioimmunoassay systems and reagents have been shown to be appropriate for measurement of chimpanzee LH, FSH, chorionic gonadotropin (CG), and prolactin (Howland et al., 1971; Reyes et al., 1975). Dilutions of chimpanzee serum from differing physiologic states were in all cases parallel to human standards. Thus, chimpanzee gonadotropins appear to be immunologically more closely related to their human counterparts than gonadotropins from other primate species. Specific steroid radioimmunoassays for estrone, estradiol-17β, estriol, progesterone, and testosterone have also been validated for use in chimpanzees (Reyes et al., 1975).

Serum menstrual cycle hormone patterns. The availability of radio-immunoassays and the development of techniques for frequent blood sample collection from unrestrained chimpanzees have allowed us to accumulate data that show that hormone patterns of the chimpanzee and human menstrual cycle are remarkably similar. Serum levels of chimpanzee LH, FSH, progesterone, and estradiol-17β during the menstrual cycle all fall within or near the normal human range (Figs. 1–4) when the days of the cycle are numbered from the midcycle LH peak. In contrast to chimpanzees and humans, in which the midcycle LH peak may extend over several days, the midcycle peak in rhesus monkeys lasts for only 1–2 days. Although the patterns of estradiol-17β concentrations are similar during rhesus, chimpanzee, and human menstrual cycles, levels of estradiol-17β are lower in rhesus monkeys. Since gonadotropin concentrations are expressed in terms of a

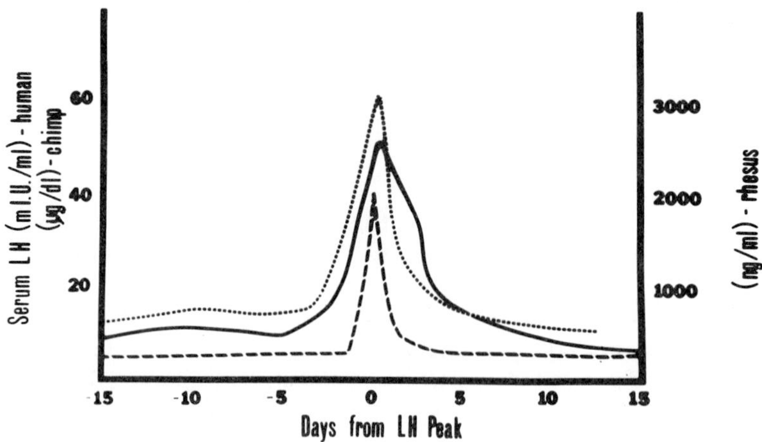

FIGURE 1.   Serum levels of LH during rhesus (– – – –), chimpanzee (————), and human (· · · · · ·) menstrual cycles. Data redrawn from Abraham et al. (1972), Faiman and Ryan (1967), Faiman et al. (1975), Howland et al. (1971), and Reyes et al. (1975).

different standard for each species, it is not possible to directly compare absolute serum LH and FSH levels among the three species.

Pregnancy.   Of the primate species studied in detail, only the chimpanzee has patterns of serum gonadotropins and steroids that are like those of humans during pregnancy. The first changes indicative of pregnancy in the

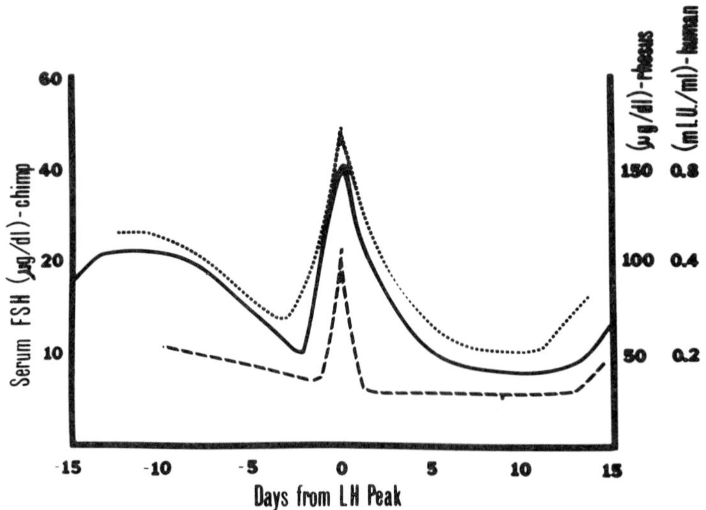

FIGURE 2.   Serum levels of FSH during rhesus (– – – –), chimpanzee (————), and human (· · · · · ·) menstrual cycles. Data redrawn from sources cited in Fig. 1.

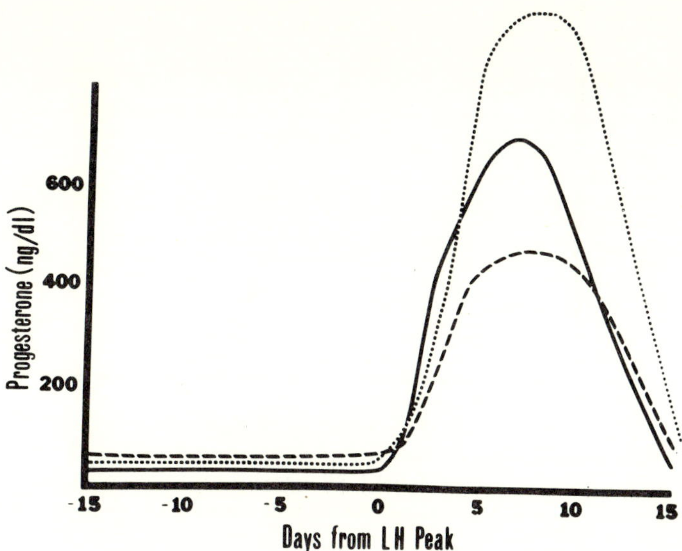

FIGURE 3. Serum concentrations of progesterone during menstrual cycles of rhesus monkeys (— — —), chimpanzees (———), and humans (- - - -). Data redrawn from Johansson (1969), Kirton et al. (1970), Mishell et al. (1971), Neill et al. (1967a), Niswender and Spies (1973), and Reyes et al. (1975).

chimpanzee occur about the tenth day after the midcycle LH peak. Levels of chorionic gonadotropin, estradiol, and estrone begin to rise and FSH falls to basal levels (Reyes et al., 1975). These changes imply that, similar to humans, implantation occurs at about this time in the chimpanzee. The high levels of progesterone seen in a few animals 5 days after the LH peak suggest that the embryo may be providing a luteotropic stimulus before implantation.

By the 15th day of pregnancy, LH–CG levels (these hormones mutually cross-react in their radioimmunoassay systems) surpass levels seen during the preovulatory LH peak (Fig. 5). CG reaches maximal levels between the 40th and 60th day of pregnancy when measured by radioimmunoassay (Reyes et al., 1975) or by bioassay (Nixon et al., 1972). Assuming a one-to-one cross-reaction in the assay, levels of CG at this time are similar to those seen in humans (Erterl et al., 1969; Faiman et al., 1968; Marshall et al., 1968; Tulchinsky and Hobel, 1973). Following this peak CG levels fall to a nadir at about 150 days and then rise gradually to term. Although the patterns are the same, the absolute levels of chimpanzee CG during the last half of pregnancy are below those of humans. In contrast, CG in rhesus monkeys peaks at far lower levels at about day 25 and is nondetectable by day 40 (Hobson et al., 1975).

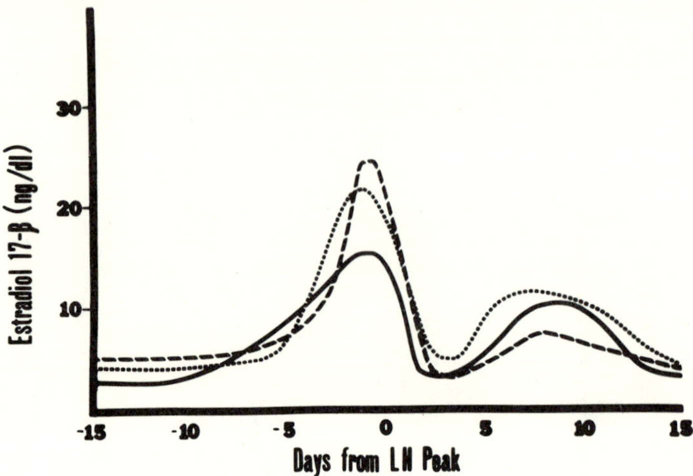

FIGURE 4. Serum concentrations of estradiol-17β during menstrual cycles of rhesus monkeys (— — —), chimpanzees (———), and humans (· · · ·). Data redrawn from Abraham et al. (1972), Baird and Guevara (1969), Hotchkiss et al. (1971), Mishell et al. (1971), and Reyes et al. (1975).

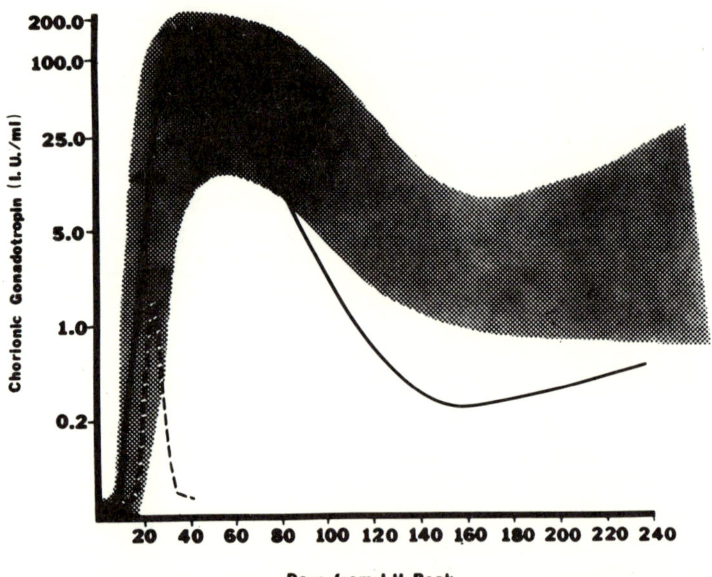

FIGURE 5. Serum concentrations of chorionic gonadotropin during pregnancy in rhesus monkeys (— — — —) and chimpanzees (———). The shaded area represents the normal range of human values. Values are expressed on a logarithmic scale. Data redrawn from Erterl et al. (1969), Faiman et al. (1968), Hobson et al. (1975), Marshall et al. (1968), Mishell et al. (1973), Reyes et al. (1975), and Tulchinsky and Hobel (1973).

The remarkably different patterns of serum estrogens and progesterone in humans and rhesus monkeys (Figs. 6 and 7) suggest that mechanisms controlling the endocrinology of pregnancy in these two species may differ. Notable among these differences is the lack of measurable levels of estriol during pregnancy in rhesus monkeys (Hodgen et al., 1972; Knobil, 1972). We were pleased, therefore, to find that serum patterns of progesterone, estradiol, estrone, and estriol in the chimpanzee are remarkably similar to those observed in humans. The estrogens rise rapidly beginning about day 10, plateau from days 20 to 80, and then rise steadily until term. Serum progesterone concentrations tend to rise coincident with the CG peak and then decline slightly at about 80 days. Progesterone begins to rise again around 140 days and reaches maximal values at term. Progesterone levels above 6 ng/ml are seldom seen in pregnant rhesus monkeys (Bosu et al., 1974; Neill et al., 1969), but in chimpanzees and humans levels often exceed

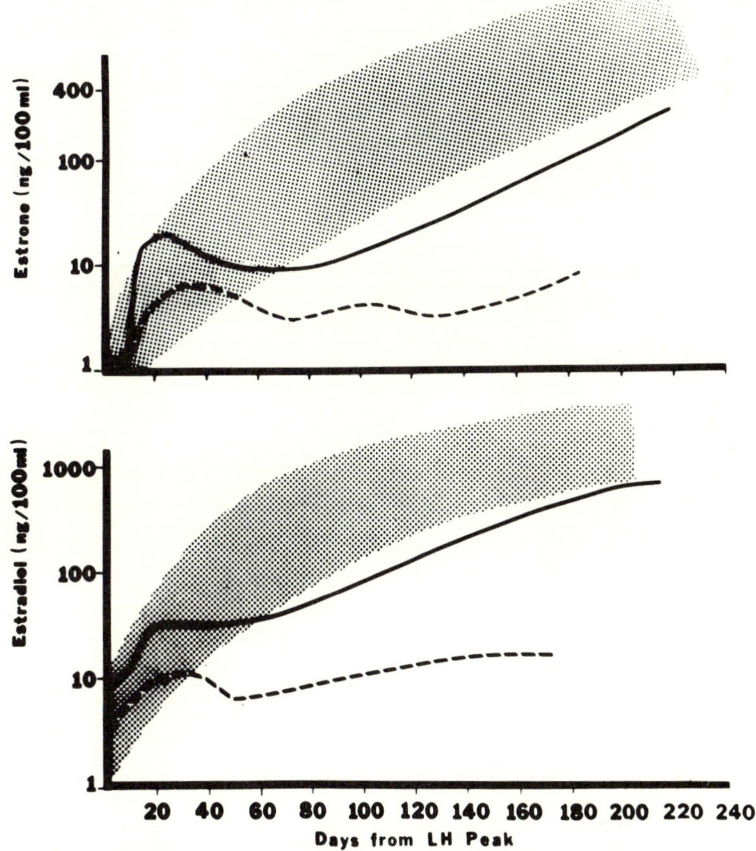

FIGURE 6. Serum concentrations of esterone and estradiol during pregnancy in rhesus monkeys (– – – –) and chimpanzees (———). The shaded area represents the normal range of human values. Values are expressed on a logarithmic scale. Data redrawn from Bosu et al. (1973), Mishell et al. (1973), Reyes et al. (1975), Tulchinsky and Hobel (1973), and Tulchinsky et al. (1972).

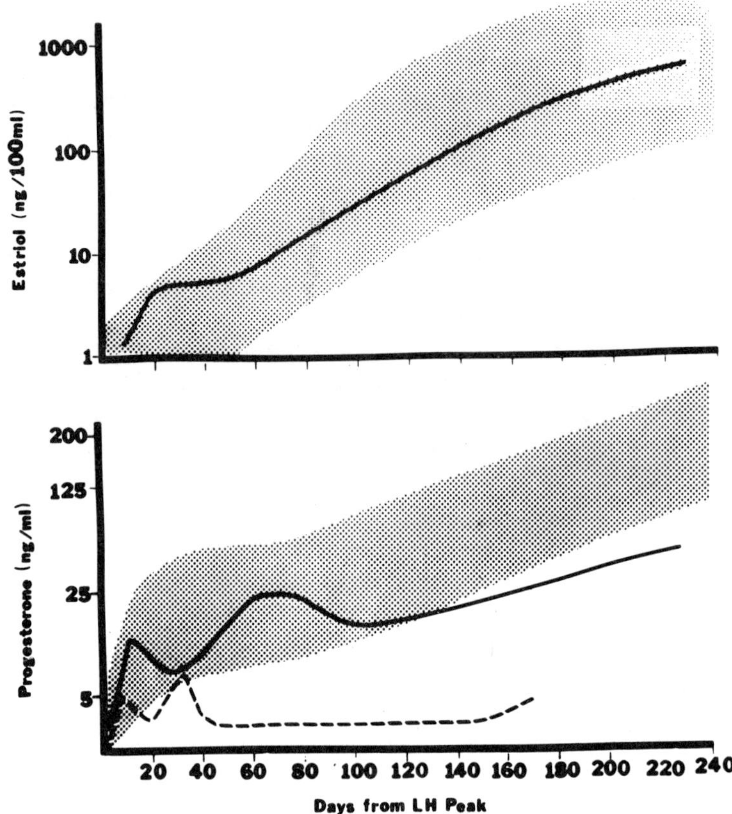

FIGURE 7. Serum levels of estriol and progesterone during pregnancy in rhesus monkeys (– – – –) and chimpanzees (———). The shaded area represents the normal range of human values. Values are expressed on a logarithmic scale. Data redrawn from Bosu et al. (1974), Loriaux et al. (1972), Mishell et al. (1973), Neill et al. (1969), Reyes et al. (1975), Tulchinsky and Abraham (1971), Tulchinsky and Hobel (1973), and Tulchinsky et al. (1972). Rhesus monkeys do not have measurable levels of serum estriol during pregnancy (Hodgen et al., 1972; Knobil, 1972).

100 mg/ml (Mishell et al., 1973; Reyes et al., 1975; Tulchinsky and Hobel, 1973).

Preliminary data indicate that no apparent pattern of prolactin concentrations is associated with the menstrual cycle in chimpanzees, but not unlike humans, levels rise throughout pregnancy (Reyes et al., 1975). Unlike the rhesus monkey, but similar to humans, chimpanzees do not appear to release prolactin in response to venapuncture or other stress.

CONCLUSION

Menstrual patterns of serum LH and FSH in the chimpanzee are remarkably similar to those of normal human menstrual cycles. The midcycle LH peak is longer in these two species than in the rhesus monkey or baboon (Faiman et al., 1975; Stevens et al., 1970) (Fig. 1). Serum levels of estradiol-17β and progesterone are similar in humans, chimpanzees, and rhesus monkeys with the exception that the rise in estrogen levels during the luteal phase of the cycle appears to be lower in rhesus monkeys.

While serum levels of estradiol and estrone are similar in humans and chimpanzees, levels of urinary excretion of these steroids and their metabolites are lower in chimpanzees. Although one might speculate that these differences are due to reduced production and clearance rates of estradiol and estrone in the chimpanzee, it seems more likely that the chimpanzee may excrete a greater proportion of these steroids or their metabolites in the feces.

The similarity of human and chimpanzee serum estrogen levels during pregnancy, the similarity of *in vitro* biosynthesis and metabolism of steroids by the placenta (Hobson, 1971; Ryan and Hopper, 1974), and the similarity of metabolism of exogenously administered estrone (Jirku and Layne, 1965) suggest that the chimpanzee and human have comparable fetoplacental steroidogenic systems. Since estriol is produced by the placenta from fetal precursors, its production has served as a measure of the health and progress of human pregnancies (Tulchinsky and Korenman, 1971). The lack of measurable concentrations of serum estriol in pregnant rhesus monkeys thus limits their value as a model of human pregnancy.

In spite of limited availability and relative expense of maintenance, chimpanzees appear to be an ideal model for experiments with application to human reproduction. Thus experiments in chimpanzees may help to resolve important questions of the safety of human use of hormonally active compounds. Specifically, the similarities of human and chimpanzee estrogen patterns and metabolism make it an especially appropriate model for studies of compounds with estrogenic properties. Moreover, since the endocrinology of pregnancy in humans and chimpanzees appears to be so closely related, the chimpanzee may serve as the final arbitrator in evaluating possible effects on human pregnancy.

QUESTIONS AND ANSWERS

D. M. Henricks: At what days in the cycle were you finding peak levels of estradiol?

W. Hobson: About days 15 and 21. In these data we have about 400 pg/ml in the follicular peak and 200–300 pg/ml in the luteal peak.

J. H. Clark: Have you ever seen anything to indicate that after mating but prior to implantation that the incipient chimpanzee embryo produces the chimpanzee equivalent of HCG?

W. Hobson: The high levels of steroids at this time indicate that something is being produced although we have not characterized it. Since it is very difficult to distinguish between CG and LH, we have not shown that it is CG in this case.

L. L. Doyle: Do hormone peak values always correlate with sex changes (sex skin swelling)?

W. Hobson: Qualitatively they correlate within expected limits; however, in quantitative terms, they are not well correlated.

L. L. Doyle: Have you looked at acyclic chimpanzees?

W. Hobson: Only during lactation.

H. D. Hafs: What is the fate of the corpus luteum in the chimpanzee during paturition?

W. Hobson: I do not know. Progesterone levels rapidly drop back to base-line values during this phase as they do in the monkey. In the monkey the corpus luteum of pregnancy is still present, and one does exist at this time in the chimpanzee; however, we do not know if it is a new or an old one.

H. D. Hafs: What is the origin of the estrogen on day 21 and 22? Is there something comparable to a midcycle follicle?

W. Hobson: We would suggest the corpus luteum as a source, but we have no direct evidence for that.

C. L. Chen: Have you tried a tranquilizing effect on prolactin secretion in the blood?

W. Hobson: No, we have not.

REFERENCES

Abraham, G. E., Odell, W. D., Swerdloff, R. S. and Hopper, K. J. 1972. Simultaneous radioimmunoassay of plasma FSH, LH, progesterone, 17-hydroxyprogesterone and estradiol-17-β during the menstrual cycle. *Clin. Endocrinol.* 34:312–318.

Baird, D. I. and Guevara, A. 1969. Concentration of unconjugated estrone and estradiol in peripheral plasma in nonpregnant women throughout the menstrual cycle, castrate and post-menopausal women and in men. *J. Clin. Endocrinol. Metab.* 29:149–156.

Bosu, W. T. K., Johansson, E. D. B. and Gemzell, C. 1973. Peripheral levels of oestrogen and progesterone in pregnant rhesus monkeys treated with dexamethasone. *Acta Endocrinol.* 74:338–347.

Bosu, W. T. K., Johansson, E. D. B. and Gemzell, C. 1974. Influence of oophorectomy, luteectomy, foetal death and dexamethasone on peripheral plasma levels of oestrogens and progesterone in the pregnant *Macaca mulatta. Acta Endocrinol.* 75:601–616.

Brown, J. B. and Carey, H. M., ed. 1963. *Modern trends in reproductive physiology,* vol. 9, pp. 153–171. Washington, D.C.: Butterworths.

Elmadjian, F. and Forchielli, E. 1965. *Congress on hormonal steroids,* vol. 2, pp. 535–539. New York: Academic Press.

Erterl, N. H., Moskoritz, M. L. and Schiffer, M. A. 1969. A modification of the rapid method for the assay of plasma estriol in pregnancy: Use of unconjugated ^{3}H-estriol to correct for losses. *J. Clin. Endocrinol. Metab.* 29:1266–1268.

Faiman, C. and Ryan, R. J. 1967. Serum FSH and LH concentrations during the menstrual cycle as determined by radioimmunoassays. *J. Clin. Endocrinol. Metab.* 27:1711–1716.

Faiman, C., Ryan, R. J., Greslin, J. G. et al. 1967. Species specificity of FSH and LH as determined by radioimmunoassay. *Proc. Soc. Exp. Biol. Med.* 125;1232–1234.

Faiman, C., Ryan, R. J., Zwirek, S. J. and Rubin, M. E. 1968. Serum FSH and HCG during human pregnancy and puerperium. *J. Clin. Endocrinol. Metab.* 28:1323–1329.

Faiman, C., Stearns, E. L., Winter, J. S. D., Reyes, F. I. and Hobson, W. C. 1975. Radioimmunoassay for rhesus monkey gonadotropins. *Proc. Soc. Exp. Biol. Med.* 149:670–676.

Graham, C. E., Collins, D. C., Robinson, H. and Preedy, J. R. K. 1972. Urinary levels of estrogens and pregnanediol and plasma levels of progesterones during the menstrual cycle of the chimpanzee: Relationship to the sexual swelling. *Endocrinology* 91:13–24.

Guilloud, N. B. 1969. The breeding of apes: Experience at the Yerkes Regional Primate Research Center and a brief review of the literature. *Ann. N.Y. Acad. Sci.* 162:297–300.

Hobson, B. M. 1971. In *Advances in reproductive physiology*, ed. M. W. H. Bishop, vol. V, pp. 67–69. Great Britain: Logos Press.

Hobson, W., Faiman, C., Dougherty, W. D., Reyes, F. I. and Winter, J. S. D. 1975. Radioimmunoassay of rhesus monkey chorionic gonadotropin. *Fertil. Steril.* 26:93–97.

Hodgen, G. D., Dufau, M. L., Katt, K. J. and Tullner, W. W. 1972. Estrogens, progesterone and chorionic gonadotropin in pregnant rhesus monkeys. *Endocrinology* 91:896–900.

Hopper, B. and Tullner, W. W. 1970. Urinary estrone and plasma progesterone levels during the menstrual cycle of the rhesus monkey. *Endocrinology* 86:1225–1230.

Hotchkiss, J., Atkinson, L. E. and Knobil, E. 1971. Time course of serum estrogen and LH concentrations during the menstrual cycle of the rhesus monkey. *Endocrinology* 89:177–183.

Howland B. E., Faiman, C. and Butler, T. M. 1971. Serum levels of FSH and LH during the menstrual cycle of the chimpanzee. *Biol. Reprod.* 4:101–105.

Jeffersy, Jd'A. 1966. Metabolism of progesterone in the pigtail monkey *J. Endocrinol.* 34:387–391.

Jirku, H. and Layne, D. S. 1965. The metabolism of estrone-^{14}C in a pregnant chimpanzee. *Steroids* 5:37–44.

Johansson, E. D. B. 1969. Progesterone levels in peripheral plasma during the luteal phase of the normal human menstrual cycle. *Acta Endocrinol.* 61:592–598.

Keeling, M. E. and Roberts, J. R. 1972. Breeding and reproduction of chimpanzees. In *The chimpanzee*, vol. 5, pp. 127–152. Baltimore, Md.: Karger.

Kirton, K. T., Niswender, G. D., Midgley, A. R., Jr., Jaffee, R. B. and Forbes, A. D. 1970. Serum LH and progesterone concentration during the menstrual cycle of the rhesus monkey. *J. Clin. Endocrinol. Metab.* 30:105–110.

Knobil, E. 1972. Hormonal control of the menstrual cycle and ovulation in the rhesus monkey. *Acta Endocrinol. (Kbh.) Suppl.* 166:137–144.

Loriaux, D. L., Ruder, H. J., Knab, D. R. and Lipsett, M. B. 1972. Estrone sulfate, estrone, estradiol, and estriol plasma levels in human pregnancy. *J. Clin. Endocrinol. Metab.* 35:887–891.

Marshall, J. R., Hammond, C. B., Ross, G. T., Jacobson, A., Rayford, P. and Odell, W. D. 1968. Plasma and urinary chorionic gonadotropin during early human pregnancy. *Obstet. Gynecol.* 32:760–764.

Mishell, D. R., Nakamura, R. M., Crosignani, P. D., Stone, S., Karma, K., Nagata, Y. and Thorneycroft, I. H. 1971. Serum gonadotropin and steroid patterns during the normal menstrual cycle. *Am. J. Obstet. Gynecol.* 111:60–65.

Mishell, D. R., Thorneycroft, I. H., Nagata, Y., Murata, T. and Nakamura, R. M. 1973. Serum gonadotropin and steroid patterns in early human gestation. *Am. J. Obstet. Gynecol.* 117:631–642.

Neill, J. D., Johansson, E. D. B. and Knobil, E. 1967a. Levels of progesterone in peripheral plasma during the menstrual cycle of the rhesus monkey. *Endocrinology* 81:1161–1164.

Neill, J. D., Peckham, W. D. and Knobil, E. 1967b. Apparent absence of immunological cross-reactivity between human and simian gonadotropic hormones as determined by radioimmunoassay. *Nature* 213:1014–1015.

Neill, J. D., Johansson, E. D. B. and Knobil, E. 1969. Patterns of circulating progesterone concentrations during the fertile menstrual cycle and the remainder of gestation in the rhesus monkey. *Endocrinology* 84:45–48.

Niswender, G. D. and Spies, H. G. 1973. Serum levels of LH, FSH and progesterone throughout the menstrual cycle of rhesus monkeys. *J. Clin. Endocrinol. Metab.* 37:326–328.

Nixon, W. E., Hodgen, G. D., Neimann, W. H., Ross, G. T. and Tullner, W. W. 1972. Urinary chorionic gonadotropin in middle and late pregnancy in the chimpanzee. *Endocrinology* 90:1105–1112.

Plant, T. M., James, V. H. T. and Michael, R. P. 1969. Metabolism of (4-^{14}C)-progesterone in the rhesus monkey. *J. Endocrinol.* 43:493–494.

Reyes, F. I., Winter, J. S. D., Faiman, C. and Hobson, W. 1975. Serial serum levels of gonadotropins, prolactin and sex steroids in the non-pregnant and pregnant chimpanzee. *Endocrinology* 96:1447–1455.

Ryan, K. J. and Hopper, B. R. 1974. Placental biosynthesis and metabolism of steroid hormones in primates. *Contrib. Primat.* 3:258–274.

Stevens, V. C., Sparks, S. J. and Powell, J. E. 1970. Levels of estrogens, progesterone and LH during the menstrual cycle of the baboon. *Endocrinology* 87:658–666.

Tulchinsky, D. and Abraham, G. E. 1971. Radioimmunoassay of plasma estriol. *J. Clin. Endocrinol. Metab.* 33:775–782.

Tulchinsky, D. and Hobel, C. J. 1973. Plasma human chorionic gonadotropin, estrone, estradiol, estriol, progesterone and 17-hydroxyprogesterone in human pregnancy. *Am. J. Obstet. Gynecol.* 117:884–893.

Tulchinsky, D. and Korenman, S. G. 1971. The plasma estradiol as an index of fetoplacental function. *J. Clin. Invest.* 50:1490–1497.

Tulchinsky, D., Hobel, C. J., Yeager, E. and Marshall, J. R. 1972. Plasma estrone, estradiol, estriol, progesterone and 17-hydroxyprogesterone in human pregnancy. *Am. J. Obstet. Gynecol.* 112:1095–1100.

Received June 4, 1975
Accepted June 27, 1975

SOME ASSAY RESTRICTIONS ON INFERENCES MADE FROM DETERMINING HORMONES IN HORSES, COWS, AND THEIR FETUSES

H. D. Hafs

Department of Dairy Science, Michigan State University,
East Lansing, Michigan

Often in developing hormone assays, hormones that may interfere with the assay by cross-reaction are not available for testing the validity of the assay. For example, horse TSH was unavailable to test for cross-reaction in an LH radioimmunoassay (RIA). The authors devised an indirect means of accomplishing the same goal, and the evidence from the indirect test of cross-reaction was at least as persuasive as a direct test might have been. Other examples are given of experiments where extensive effort was devoted to validation of steroid RIA, but there were substantial quantitative differences in the results among experiments and among laboratories. Differences of this kind probably would be intolerable in an assay used to monitor hormone residues in food-producing animals.

INTRODUCTION

During the past decade one overall objective of some of our research was to describe the changes in hormones that normally occur during sexual differentiation of the fetus, at birth, during puberty, during the estrous cycle, during pregnancy, and at parturition. We speculated that knowledge of normal changes would provide clues that might lead to ways of managing hormones to control reproductive events. During the course of our research, we encountered some discrepancies that, I believe, may be pertinent to the proposed NITR estrogen research program. Second, the published data reveal that we have only begun to uncover the complexity of the endocrine controls of reproduction, including growth and galactopoesis, suggesting that we would be well advised not to oversimplify implications on the relationship of hormones to mammary tumorigenesis.

Data are presented here to illustrate some discrepancies in hormone assays and the variety of hormones that may be involved with mammary growth and galactopoesis.

This research was supported in part by USPHS research grant HD-06720 and HD-06948 and is published with the approval of the Director of the Agricultural Experiment Station as paper no. 7149.

Requests for reprints should be sent to H. D. Hafs, Department of Dairy Science, Michigan State University, East Lansing, Michigan 48824.

Journal of Toxicology and Environmental Health, 1:669–679, 1976

RESULTS AND DISCUSSION

Ovulatory Surge of LH in Horses

A relatively brief pulse of blood LH persists for only 8–12 hr before ovulation in most species. However, Noden et al. (1974) have described an ovulatory surge of LH that persisted about 10 days in mares; blood LH began to rise before onset of estrus and continued to increase for about 5 days until after ovulation, which occurred 1.5 days before the end of estrus. Then LH declined symmetrically with the increase (Fig. 1). The reason for the prolonged ovulatory surge of LH in mares is a mystery. Others (Pattison et al., 1972; Whitmore et al., 1973) agree to the long duration.

The purpose in raising this biological enigma here is that assay of equine LH presented some methodologic problems. First, universally accepted standardized equine LH was not available. As a consequence, Noden et al. (1974), Pattison et al. (1972), and Whitmore et al. (1973) used different LH standards, thus compounding the task of making quantitative comparisons among the three publications. Second, highly purified equine pituitary hormones were not available. Therefore, it was not possible to test directly whether the LH assay cross-reacted with the other pituitary hormones. However, often one can devise indirect methods to test assay specificity, and these methods may be desirable even when more direct tests are available.

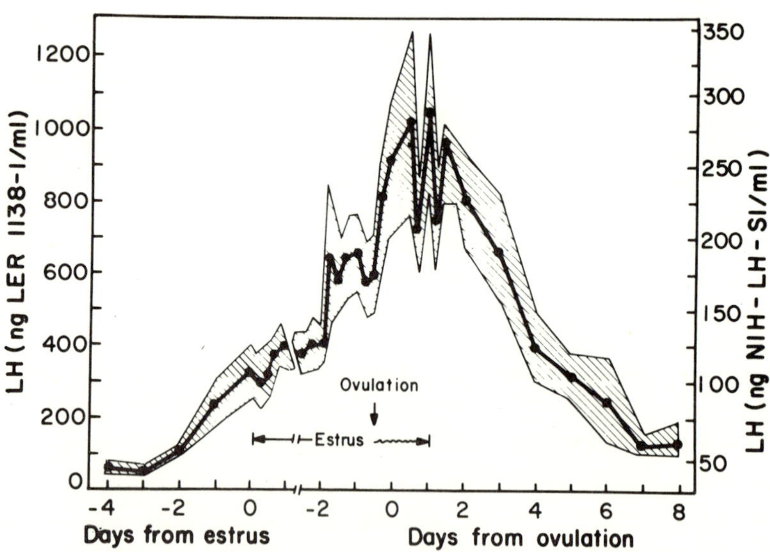

FIGURE 1. Blood serum LH during control estrous cycles in six mares. Data from Noden et al. (1974).

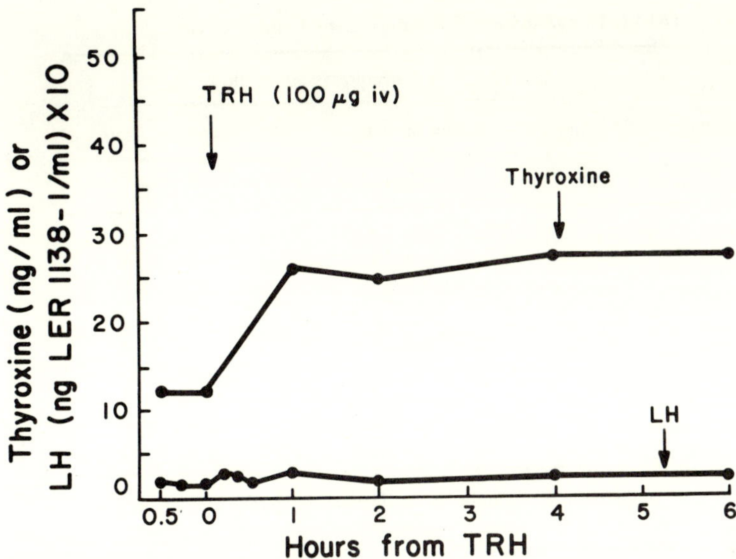

FIGURE 2. Blood serum LH and thyroxine after iv administration of 100 μg thyrotropin-releasing hormone to two pregnant mares. Data from Noden et al. (1974).

For example, since purified equine TSH was not available to test for cross-reaction in the LH radioimmunoassay (RIA), Noden et al. (1974) measured LH after administration of thyrotropin-releasing hormone (TRH) to pregnant mares (Fig. 2). The pregnant mares had suppressed blood LH due to high blood progesterone; the TRH injection caused release of TSH as reflected in the increased blood thyroxine. But LH did not increase after injection of TRH. This constitutes persuasive evidence that the assay for LH did not detect TSH, even under conditions of the largest TSH:LH ratios likely to be encountered normally in horse blood. Nevertheless, a direct test for cross-reaction of purified TSH, as well as of the other pituitary hormones, would provide additional confidence on the validity of this LH assay.

Estradiol during the Estrous Cycle

Wettemann et al. (1972) described the changes in jugular blood estradiol during the bovine estrous cycle, as measured by RIA and competitive-protein-binding (CPB) methods. The results of the two independent methods (Table 1) *qualitatively* agreed well with each other, as well as with comparable data published by others. Both methods were sufficiently sensitive to detect the quantities of estradiol in blood during the estrous cycle. Regardless of author or assay method, we concluded that blood estradiol begins to rise 2 or 3 days before estrus, peaks near the onset of estrus, and then falls to low values that persist from metestrus through diestrus. This kind of qualitative change information

TABLE 1. Estradiol in Cows during the Estrous Cycle[a]

Day of cycle	Protein binding		Radioimmunoassay	
	n	pg/ml	n	pg/ml
− 3	3	5.9 ± 2.9[b]	4	4.8 ± 1.2
− 2	4	9.0 ± 2.2	8	7.7 ± 1.7
− 1	7	11.6 ± 2.0	11	8.7 ± 1.1
−0.5	3	10.5 ± 2.2	5	9.7 ± 2.2
0	7	12.3 ± 3.8	9	8.4 ± 1.6
2	6	5.7 ± 1.5	10	3.0 ± 0.9
4	4	8.5 ± 2.7	10	3.9 ± 0.4
7	4	7.4 ± 1.9	10	3.8 ± 0.6
11	4	7.5 ± 1.9	10	3.6 ± 0.4

[a]Data from Wettemann et al. (1972). Different sets of sera were used for the two assay methods.
[b]Values are means ± SE.

seemed adequate for making inferences about the relationships of estradiol secretion to luteolysis, to the ovulatory surge of LH, to luteinization of granulosa cells, to ovulation, and to estrual behavior.

Wettemann et al. (1972) showed that their estradiol RIA and CPB assays both cross-reacted with estrone and estriol. Consequently, to obtain the data in Table 1, they isolated estradiol from estrone and estriol by column chromatography. Even after chromatographic isolation of estradiol, the CPB assay resulted in appreciably higher values than the RIA. Most of this difference probably was caused by higher background blank values from the CPB assay. Evidently, unidentified factors influenced the CPB assay more than the RIA. In addition, the repeatability of the CPB assay was inferior to that of the RIA, as reflected by larger standard errors for the CPB assay in Table 1. Based on these observations, the RIA is the method of choice, even for the experimental implications published by Wettemann et al. (1972) because of its greater specificity and its greater precision. In my opinion, given the potential consequences of information from an assay widely used to monitor estrogen residue in food, the assay must be at least as specific, sensitive, and precise as the RIA in Table 1. In other words, while assays that are only qualitatively precise may suffice for some kinds of inferences, an assay used to monitor residues in food must be quantitatively precise as well.

Maternal and Fetal Estradiol and Testosterone

Radioimmunoassays had been validated for assay of estradiol and progesterone in cows during the estrous cycle, during pregnancy, and around parturition (Wettemann et al., 1972; Smith et al., 1973), as well as for testosterone in pubertal and adult bulls (Smith and Hafs, 1973). Thus, one might be inclined to use these same RIA in experiments to compare gonadal steroids in blood of cows with that of their fetuses. However, RIA that were validated for use in cow blood may not be used for fetal blood without further validation, because fetal blood could contain interfering substances not found in cow blood. Furthermore, in the absence of proof of identity of the fetal and adult protein hormones, structural differences may not permit assay of the fetal hormone using an assay based on antiserum prepared against the adult hormone.

Mongkonpunya et al. (1975) concluded that male fetuses had tenfold higher testosterone and twofold higher androstenedione than female fetuses, male fetal testosterone decreased from adult values at 90 days of gestation to 0.3 ng/ml near parturition, and maternal blood androgens increased during pregnancy in cows with male fetuses but not in cows with female fetuses (Tables 2 and 3).

Lin et al. (1974) concluded that the placenta probably was no barrier to estradiol because maternal and fetal concentrations of estradiol were similar at the end of each trimester of pregnancy, estrone was about fivefold higher in maternal blood than in fetal blood at 180 and 270 days of gestation (Figs. 3 and 4), and the placenta evidently excluded progesterone because while maternal blood progesterone ranged from 5.1 to 9.5 ng/ml, fetal progesterone rarely exceeded 0.2 ng/ml. Conclusions such as these are justifiable only after rigorous validations of the assays at least as extensive as Mongkonpunya et al. (1975) published for testosterone and androstenedione. They could not assume that assays validated in bull blood were equally valid in fetal blood.

Endocrine Corrolates with Galactopoesis

There is a tendency for some to implicate a single hormone in induction of mammary tumors. But several hormones are secreted in

TABLE 2. Serum Testosterone and Androstenedione Concentrations in Bovine Fetuses[a]

Hormone	Sex	Umbilical vessel	Days gestation 90[b]	180	260
Testosterone	Male	Artery	2.76 ± 0.48	0.70 ± 0.16	0.16 ± 0.06
		Vein	—	0.50 ± 0.04	0.26 ± 0.07
	Female	Artery	0.22 ± 0.09	0.07 ± 0.02	0.06 ± 0.03
		Vein	—	0.10 ± 0.03	0.02 ± 0.01
Androstenedione	Male	Artery	2.68 ± 1.11	1.01 ± 0.51	0.52 ± 0.10
		Vein	—	2.20 ± 0.94	1.15 ± 0.47
	Female	Artery	1.86 ± 0.51	0.55 ± 0.06	0.21 ± 0.05
		Vein	—	0.93 ± 0.28	0.44 ± 0.10

[a]Data from Mongkonpunya et al. (1975). Values are means ± SE ($n = 5$–7), expressed in ng/ml.
[b]Samples for cardiac puncture for 90-day fetuses.

TABLE 3. Serum Concentrations of Testosterone and Androstenedione in the Uterine Vessels from Pregnant Heifers[a]

Hormone	Sex of fetus	Uterine vessel	Days gestation		
			90	180	260
Testosterone	Male	Artery	0.35 ± 0.05	0.59 ± 0.06	0.74 ± 0.10
		Vein	0.51 ± 0.03	0.86 ± 0.07	0.82 ± 0.11
	Female	Artery	0.13 ± 0.02	0.28 ± 0.12	0.37 ± 0.10
		Vein	0.14 ± 0.02	0.30 ± 0.08	0.31 ± 0.10
Androstenedione	Male	Artery	0.25 ± 0.13	1.00 ± 0.21	1.39 ± 0.35
		Vein	2.02 ± 0.48	3.96 ± 1.00	3.20 ± 0.78
	Female	Artery	1.00 ± 0.13	0.85 ± 0.14	1.15 ± 0.19
		Vein	4.55 ± 1.73	2.82 ± 1.14	2.35 ± 0.56

[a]Data from Mongkonpunya et al. (1975). Values are means ± SE ($n = 5$–7), expressed in ng/ml.

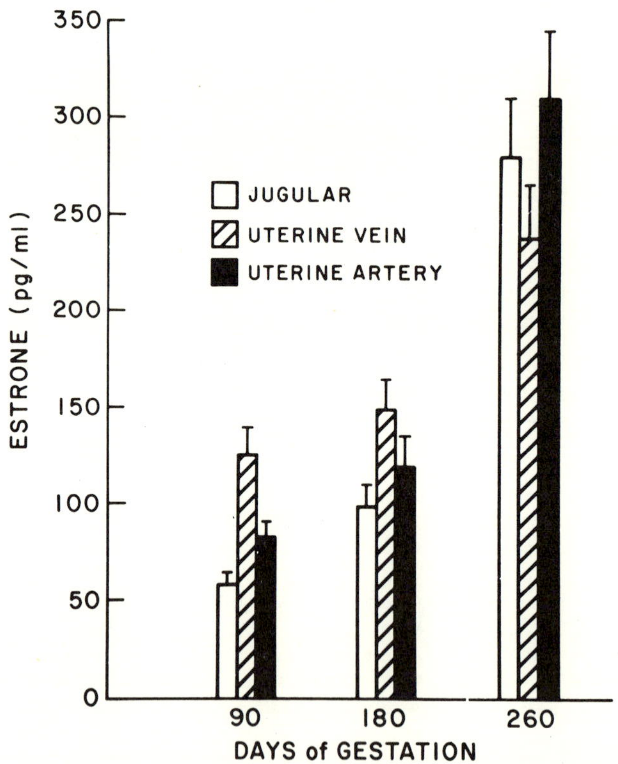

FIGURE 3. Blood serum estrone in cows bearing female fetuses. Data from Lin et al. (1974).

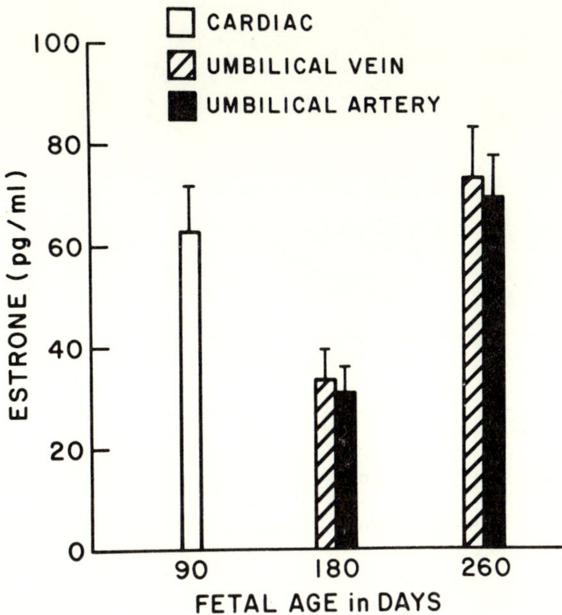

FIGURE 4. Blood serum estrone in bovine female fetuses. Data from Lin et al. (1974).

increased quantities during late pregnancy and around parturition and thus may be involved in mammary growth and lactogenesis. In cows, for example, both Ingalls et al. (1973) and Edgerton and Hafs (1973) described the dramatic elevations of prolactin and growth hormone that occur during the week before parturition; Kumaresan and Turner (1965) showed that insulin is required for mammary growth. All three hormones probably are involved in mammary growth or galactopoesis.

Edgerton and Hafs (1973) and Smith et al. (1973) described changes in progesterone, estradiol, estrone, and glucocorticoid late in pregnancy and around parturition. During the last month of pregnancy, blood estradiol increased to values at least 50-fold higher than those that typify estrus, and concentrations of estrone (Fig. 5) were about eightfold greater than estradiol. Both estradiol and estrone began to fall about 2 days before parturition, coincidentally with the fall in blood progesterone (Fig. 5). The high progesterone and the dramatically elevated estrogen during the month before parturition are thought to be the principal endocrine controls responsible for growth of mammary ducts and alveoli, at least those occurring before the major changes in prolactin and growth hormone during the week before parturition. The marked surge of glucocorticoid, which normally occurs at parturition (Fig. 5), probably is involved in lactogenesis more than in mammary growth.

How pituitary, pancreatic, and adrenal hormones participate with estrogens, progesterone, and glucocorticoid in causing mammary growth

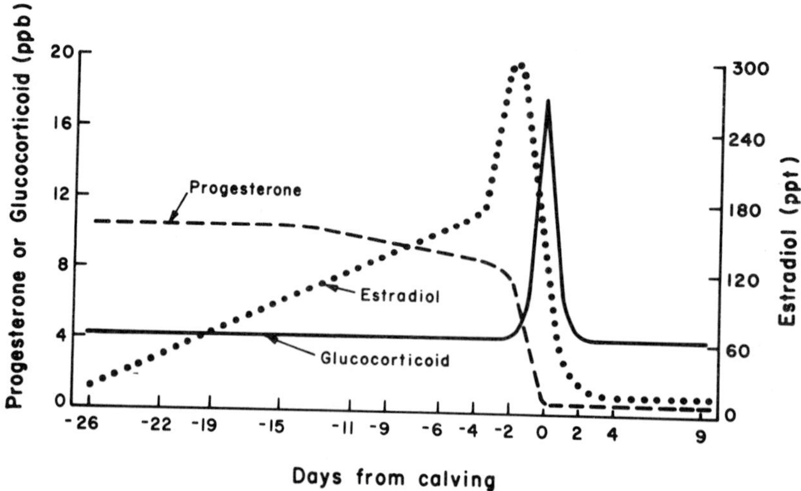

FIGURE 5. Idealized summary of changes in progesterone, estradiol, and glucocorticoid in cows near parturition. Data from Edgerton and Hafs (1973) and Smith et al. (1973).

and lactogenesis is not known with certainty. In my opinion, however, inferences between changes in any one hormone and tumorigenesis may be misleading, unless the other mammogenic hormones are considered simultaneously.

When compared with data published by others, the data in Fig. 5 reveal some potential assay discrepancies. We reported a decided fall in estrone and estradiol during the 2 days before parturition, in agreement with data published by Robinson et al. (1971). First the values from Robinson et al. (1971), derived from a fluorometric assay, were nearly threefold higher than ours from RIA. Second, in contrast to our data, Hendricks et al. (1972) found blood total estrogens peaked at parturition. This suggests that some immunoreactive substances other than estradiol and estrone were measured in their RIA, but not in ours. In my opinion, quantitative and qualitative differences such as those between our estrogen data and those published by both Robinson et al. and Hendricks et al. would be intolerable for any assay used to monitor estrogen in food.

CONCLUSION

The principal informational purposes of the experiments presented above are self-explanatory. The data derived from our research with cattle appear to be informationally logical, consistent with each other, meaningful in terms of understanding endocrine control of reproductive events, and largely consistent with comparable data from other species. These experiments were designed principally to make qualitative inferences from changes in concentrations of hormones during fetal growth, birth, puberty, the estrous cycle, and pregnancy. The assays used generated data that were qualitatively useful, but the hormone data from our experiments did not agree quantitatively with all comparable data from other laboratories.

REFERENCES

Edgerton, L. A. and Hafs, H. D. 1973. Serum luteinizing hormone, prolactin, glucocorticoid and progestin in dairy cows from calving to gestation. *J. Dairy Sci.* 56:451–458.

Hendricks, E. M., Dickey, J. F., Hill, J. R. and Johnston, W. E. 1972. Plasma estrogen and progesterone levels after mating and during late pregnancy and postpartum in cows. *Endocrinology* 90:1336–1342.

Ingalls, W. G., Convey, E. M. and Hafs, H. D. 1973 Bovine serum LH, GH and prolactin during late pregnancy, parturition and early lactation. *Proc. Soc. Exp. Biol. Med.* 143:161–164.

Kumaresan, P. and Turner, C. W. 1965. Effect of insulin and alloxan on mammary gland growth in rats. *J. Dairy Sci.* 48:1378–1381.

Lin, Y. C., Oxender, W. D. and Hafs, H. D. 1974. Estrogens and progesterone in the fetus and cow. *J. Anim. Sci.* 39:217.

Mongkonpunya, K., Lin, Y. C., Noden, P. A., Oxender, W. D. and Hafs, H. D. 1975. Androgens in the bovine fetus and dam. *Proc. Soc. Exp. Biol. Med.* 148:489–493.

Noden, P. A., Oxender, W. D. and Hafs, H. D. 1974. Estrus, ovulation, progesterone and luteinizing hormone after prostaglandin $F_{2\alpha}$ in mares. *Proc. Soc. Exp. Biol. Med.* 145:145–150.

Pattison, M. L., Chen, C. L. and King, S. L. 1972. Determination of LH and estradiol-17β surge with reference to the time of ovulation in mares. *Biol. Reprod.* 7:136.

Robinson, R., Anastassiadis, P. A. and Common, R. H. 1971. Estrone concentrations in the peripheral blood of pregnant cows. II. Values around parturition. *J. Dairy Sci.* 54:1832–1834.

Smith, O. W. and Hafs, H. D. 1973. Competitive protein binding and radioimmunoassay for testosterone in bulls and rabbits; blood serum testosterone after injection of LH or prolactin in rabbits. *Proc. Soc. Exp. Biol. Med.* 142:804–810.

Smith, V. G., Edgerton, L. A., Hafs, H. D. and Convey, E. M. 1973. Bovine serum estrogens, progestins, and glucocorticoids during late pregnancy, parturition and early lactation. *J. Anim. Sci.* 36:391–396.

Wettemann, R. P., Hafs, H. D., Edgerton, L. A. and Swanson, L. V. 1972. Estradiol and progesterone in blood serum during the bovine estrous cycle. *J. Anim. Sci.* 34:1020–1024.

Whitmore, H. L., Wentworth, B. C. and Ginther, O. J. 1973. Circulating concentrations of luteinizing hormone during estrous cycles of mares as determined by radioimmunoassay. *Am. J. Vet. Res.* 34:631–636.

Received June 4, 1975
Accepted June 27, 1975

REFLECTIONS

The following are a few points and questions about the previous presentations, by way of initiating a discussion.

1. How heavily should we rely on mouse mammary tumorigenesis as an index of carcinogenicity in man? Does anyone seriously doubt that estrogen will increase the incidence or enhance the growth of mammary

tumors in C3H mice? Is it possible that mice are not the best model? Inconsistencies between the endocrine control of mouse mammary gland and human mammary gland development suggest we should consider other animal models. For example, pregnancy in mice is known to increase the incidence of mammary tumors; in humans, however, it does just the opposite. Although mice are a valuable animal model, I am delighted to see that other models are being considered.

2. I am especially intrigued by D. Medina's observation that neonatal estrogen treatment increases the incidence of tumors induced by substances such as DMBA later in life. Also, clearly his morphological model of alveolar and ductal hyperplasia offers perhaps an ideal rapid screening technique of assaying carcinogenesis.

3. Notwithstanding how uncomfortable his data made some of us who rely on blood serum hormone data, studies such as those described by J. H. Clark have gone a long way toward uncovering the mechanism of action of steroid hormones. These techniques will be extremely useful in elucidating the mechanism of estrogen involvement in carcinogenesis.

4. From the point of view of understanding the potential hazards of steroid hormones, I am ecstatic to see the data presented by L. J. Fischer and C. C. Kaltenbach on the distribution of estrogens in humans and food-producing animals. For the first time, to my knowledge, Kaltenbach has shown the overall distribution and fate of exogenous estradiol in a food-producing animal. Also for the first time, Fischer has outlined the fate of ingested stilbestrol in humans. These data should be followed by similar studies in greater detail and by surveys of selected populations of people who for one reason or another may not eat meat. In the final analysis, answers to many of these questions must come from studies with humans.

5. The general endocrinologic similarity of the chimpanzee to humans as presented by W. Hobson is impressive when one consideres a particular hormone and its specific responses. Here, the most appropriate animal model could well turn out to be a cow, a mouse, or a pony mare. In other words, for any specific problem area, we should find an animal model that resembles humans as closely as possible.

6. In my view, this proposed estrogen research program is one of the most ambitious proposals that I have ever seen. It leaves no significant method or philosophical avenue of approach omitted as far as I could find. It proposes to use the most modern methods in a good mix with some of the older tried and tested ones. Although the rationale is clear, the ideas logical, and the approaches feasible, I am not competent to judge if the program is realistic because of the technical expertise necessary to complete the job.

If I have any reservation about the overall proposal, it is the heavy reliance on the mouse model for carcinogenesis. There are some important general endocrine differences between rodents and man, as noted above. Several hormones are intimately involved in mammary growth and galactopoesis, and these same hormones most likely play a large part in mammary tumors.

7. It seems to me that the estrogen research program arose out of stilbestrol contamination of human food from ruminants and whether it may be harmful to man. It seems, therefore, that studies in the estrogen research program should emphasize the determination of hormone contaminants found in food-producing animals and the potential hazards of these contaminants to man.

8. Finally, assaying hormones and determining the extent of hormonal contamination of food are problem areas that resemble those in assaying for gold. The same characteristics are required for any desirable assay. Whether for gold or stilbestrol the method must be sensitive, specific, repeatable, and inexpensive.

OBITUARY

GORDON M. TOMKINS

The editors of the *Journal of Toxicology and Environmental Health* acknowledge with sorrow the death of Dr. Gordon M. Tomkins, highly respected member of our Editorial Board and of the entire scientific community.

Responsible for the elucidation of the process through which cholesterol is synthesized by the liver, Dr. Tomkins was considered a major authority on hormone action and development. After graduating from Harvard Medical School in 1949, he attended the University of California at Berkeley and received his doctorate in biochemistry. His training as a physicist and biochemist resulted in his becoming recognized by his colleagues as a man with extremely broad knowledge and incisiveness.

Dr. Tomkins was chief of the Laboratory of Molecular Biology in the National Institute of Arthritis, Metabolism and Digestive Diseases at the National Institutes of Health for seven years; subsequently he became a Professor of Biochemistry and Vice Chairman of the Department of Biochemistry and Biophysics at the University of California in San Francisco, where he remained until his death.

In his last published paper, Dr. Tomkins theorized that hormones worked through a "metabolic code," a notion that, according to many scientists, could be as significant to biological principles as the genetic code is to heredity.

An accomplished musician, Dr. Tomkins often performed with jazz bands in Washington nightclubs and accompanied his wife, Millicent, a singer, in numerous classical concerts.

Dr. Tomkins' sudden death is a great loss to the scientific community, and we extend our sincerest condolences to his wife, Millicent; his daughters, Leslie and Tanya; and his mother, Mrs. Jean Tomkins.

ANNOUNCEMENTS

SYMPOSIUM ON THE HANDLING OF TOXICOLOGICAL INFORMATION, sponsored by the Toxicology Information Subcommittee of the DHEW Committee to Coordinate Toxicology and Related Programs, will be held on Thursday and Friday, May 27-28, 1976, at the Jack Masur Auditorium of the Clinical Center, National Institutes of Health, Bethesda, Maryland 20014. The number of attendees will be limited to 450. Advance registration is required and should be sent to Dr. George J. Cosmides, Toxicology Information Program, National Library of Medicine, 8600 Rockville Pike, Bethesda, Maryland 20014. Telephone (301) 496-3147; (FTS) 496-3147. There is no registration fee. The advance registration should include the following information: (1) name, address, phone number, and company or institutional affiliation; (2) a statement of the registrant's needs for toxicological information services or of involvement in the handling of toxicological information.

One of the purposes of this symposium is to provide a forum for the exchange of information among those who collect, manage, or use toxicological information in industry, academia, and government. The symposium commemorates the tenth anniversary of the President's Science Advisory Committee Report on the handling of toxicological information. On-going toxicology information activities will be surveyed in light of the recommendations in that Report, and the needs for new toxicological information services will be examined. There will be lectures and panel presentations with discussions from the floor. The program will concentrate mostly on government-sponsored information systems and services in toxicology and related fields.

Dr. Henry M. Kissman
Chairman, Toxicology Information Subcommittee
Associate Director, Specialized Information Services
National Library of Medicine

THE TERATOLOGY SOCIETY will hold its Sixteenth Annual Meeting June 20-23, 1976 at Highlands Inn, Carmel, California. For further information write Dr. Lucille S. Hurley, Department of Nutrition, University of California, Davis, California 95616.

NUTRITION AND DRUG INTERRELATIONS, a three-day international symposium sponsored by the Nutrition Foundation and the Iowa State University Nutrition Sciences Council, will be held on August 4-6, 1976, at Iowa State University. For further information write Dr. John N. Hathcock, Department of Food and Nutrition, Iowa State University, Ames, Iowa 50010.